NOISE ASSESSMENT AND CONTROL

K A MULHOLLAND AND K ATTENBOROUGH

Construction Press
London and New York

Construction Press
Longman House
Burnt Mill, Harlow, Essex, UK.

A division of Longman Group Ltd., London.

Published in the United States of America by Longman Inc, New York.

First published 1981

British Library Cataloguing in Publication Data

Mulholland, K A
  Noise assessment and control.
  1. Noise control
  I. Title
  II. Attenborough, Keith
  620.2'3   TD892

ISBN 0-86095-882-5

Printed in Great Britain by
William Clowes (Beccles) Ltd, Beccles and London.

Contents

Contents

Preface

The primary aim of this book is to provide a guide and an approach to noise control for those who are meeting the subject for the first time, in particular, members of the public who wish to tackle noise problems. However, the architect, building designer, local environmental health officer, consulting engineer and student of environmental science will also find this book useful. We will deal principally with the legal and planning aspects of noise control. The more technical aspects of noise reduction at source or at the listener are likely to be difficult for the layman, and are already adequately spelt out in a number of texts. We aim to show what specific action can be taken to control a noise nuisance.

The individual who has a noise problem is often faced with a lack of detailed knowledge of anything but his own frustration and annoyance. It is possible for him to find books which give detailed introduction to the laws pertaining to noise nuisance; however, these are being rapidly outdated by new developments. In addition, he should be aware of the planning controls on noise which are available, the units and other methods used in assessing noise nuisance, and the basis underlying these units and methods. He should be familiar, for example, with the psychological variables and social survey techniques which are used. With this information he will be in a better position to judge the strength and perhaps the weaknesses of the regulations that seek to control noise. These are among the aspects of noise that we consider in this book, and, where necessary, references for further study are given, since no one text can embrace them all fully.

Of course, we do not wish to confine our discussion merely to the sufferers. It is important for those responsible for the noise nuisance to be aware of its power as a nuisance and the steps that they can take for its alleviation. In this respect some outline of the economic analyses relating to noise nuisance and its prevention and cure will be of value and this is included.

Acknowledgements

We are grateful to the Open University for permission to reproduce Figs. 2.1, 2.2, 2.3, 4.2, 4.3, 4.4, 4.5, 4.6, 5.1, 5.2, 5.3, 8.1 and Table 5.1 from PT272 *Environmental Control and Public Health,* Units 11 and 12: 'Control of the acoustic environment', and Fig. 4.1 from PT272 *Home Experiment Handbook.*

The authors would like to thank Carol Mulholland for typing the final version of the manuscript.

1. Introduction

1.1 WHY CONTROL NOISE?

Most people tell tales of noise problems. Noise is a form of air pollution and like other forms of pollution it affects the quality of life and so can be thought of as a social cost. It is a social cost because the noise producers bear only part of the burden from the noise that is produced and for them this is outweighed easily by the benefit that accompanies the production of noise. The major part of the burden can fall upon people who are not directly party to any benefits. In effect their annoyance is benefiting the noise producer. Noise annoys, distracts, disturbs, and, when exposure to it is sufficient, noise can cause physiological effects leading to deafness. Annoyance results from interference with sleep and with speech. Noise within your home causes disturbance and loss of privacy. Distraction accompanies noise in the workplace with consequent reduction in productivity, efficiency, accuracy and safety.

As far as actual damage is concerned, however, the principal risk is damage to the human ear. Apart from noise-induced deafness, most of the effects of noise vanish with control of the noise itself. This simple fact distinguishes noise pollution from other types of pollution. Furthermore, we already have the technology to control noise in many situations – at a cost. The same cannot easily be said of pollution from the dumping of toxic wastes or from tall chimneys. Indeed, it is probable that our knowledge of how to control noise exceeds our knowledge of its physical and psychological effects. We ought to reduce this knowledge gap but at the same time we have no excuse for not solving and anticipating noise problems to the best of our ability, taking into account the social and cost factors involved.

1.2 A GENERAL METHODOLOGY OF NOISE CONTROL

A random glance at one specific week's news stories in the UK concerning noise problems might show a number of problems ranging from a village-wide disturbance due to bell-ringing, to fears of airport extensions at small domestic airports. If the protagonists involved in the bell-ringing problem have invoked the Public Health Act (1936) and the Control of Pollution Act (1974) they would seem to be well-organised and on the attack. They are a well informed cross-section of people who have realised that the first steps in any noise control procedure are identification of the source of disturbance and of appropriate standards and laws that control the degree of noise that is permitted. On the other hand the people involved near the airport seem to have a much more

1

difficult problem even with the help of the new environmental health authority powers on noise. There are few laws that control noise from airports. (We explore the problem of airport noise in Chapter 5).

Many of the steps that can be taken to control noise can be taken whatever the noise source. This fact stems from the fundamental physical laws that govern the way sound propagates and from the common geography of any noise problem which determines the areas in which control can be exerted. These areas are:

1) at the source,
2) along the path of sound between the source and receiver (listener or victim), and
3) at the receiver.

Figure 1.1 shows the flow diagram for the method of attacking any particular noise problem. Entry is at the top to the box marked 'Identify Noise Source'. It is difficult to attack a noise problem which consists of an unknown noise. Indeed, attitudes to the source or to the noise producer can have a big influence on annoyance. When human relationships are involved, the procedure divides according to whether or not the noise source is already in existence. An important consequence of this question is whether or not the person complaining about the noise has been exposed for some period of time. For example, it will not avail the purchaser of a new house to decide that it is exposed to an unacceptable level of noise after the purchase. There is then very little redress in law against the vendor since the principle of *caveat emptor* operates under which the purchaser has a duty to have been aware of the noise before purchase at a price which is assumed to reflect the existence of the noise. However, the purchaser or lessee of property can bring an action under common law against the noise maker (see Section 7.3). It is also possible that explicit statements made by the vendor before the sale was agreed could be used as a basis for redress.

Where problems arise from noise sources that are about to come into the reckoning, for example if a new motorway is to be built near housing, the occupants will want to be able to predict the likely noise level. Noises which are existing already or which have recently increased in loudness can, of course, be measured, and this produces objective data about such things as the loudness, frequency and duration of the noise. All these factors must be considered when beginning to follow up the method of Figure 1.1. Next it must be understood that measurements or predictions of noise levels alone, are useless without some means of checking them against some relevant standard. In other words, different standards are available for hearing-damage risk, local traffic noise, construction noise and acoustic privacy. Also, practices in different countries vary. In the USA for example, there are federal and state regulations as well as local noise ordinances. The relevant standard to be applied to a noise complaint, therefore, must be found and applied to the problem.

Once the standard has been identified and measurements or predictions have been made, it is necessary to ask whether the standard is satisfied. If it is, there is no problem, in the sense that there is no legal way and probably no scientific or objective way in which complaints can be justified. If you are the

Figure 1.1 A methodology of noise control

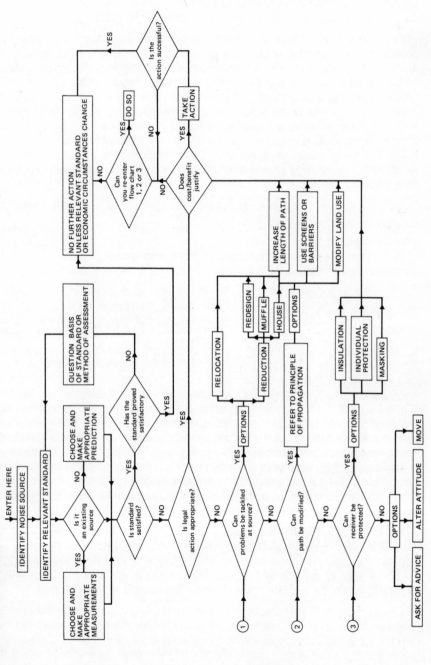

3

annoyed victim of such circumstances, then the only rational solution is to move away from the noise source. This may seem hard but very often it is the only thing that can be done without causing a greater nuisance burden to fall on other people than the one who is suffering – yourself.

If the standard is not satisfied then appropriate legal action can cause noise producers to abate the source. Legal action to abate noise nuisance is not always successful as there are well-tried defences against such prosecutions. A common defence is that of the 'best practicable means'. In the UK if the noise producer (being a business) can show that he is using the best practicable means to abate the noise nuisance, he may in fact be permitted to make a noise above an otherwise accepted standard. He can claim that it is technically impossible to reduce the noise level. Airlines are exempt from prosecution for noise nuisance although there are some controls and in the UK the Civil Aviation Authority (the Federal Aviation Authority in the USA) is responsible for their implementation.

As long as there is some method of controlling the noise the method of approach can be analysed into

1) control at source,
2) control between source and receiver (the path),
3) control at the receiver.

The most economical and often the simplest method of reducing a noise is at the source. If the source can be enclosed, treated, removed, altered or otherwise made quieter, instantly the noise problem throughout the entire area of its previous influence is reduced. Alternatively, you can plan to zone or locate a noisy process, device or area away from noise critical areas such as living quarters (or *vice versa*). Modification of the noise source is often poss-ible. Common devices are enclosures, which must include a certain amount of sound absorption, silencers for exhausts and vibration isolators or dampers. Often the cost of treating the noise at source is the minimum cost that has to be incurred in order to abate the noise. Effective noise control between source and receiver is difficult. A lot is known physically about the methods of reducing noise along paths but these very often are only partially effective. For example, if a screen is put up close to a garden near a motorway the garden itself may become quieter but the surrounding areas will not be affected. Alternatively, if the motorway is screened completely all adjoining gardens and many associ-ated residences are quietened.

So, to recapitulate, the first stage in any problem is to identify the source and to note its characteristics. These characteristics will determine the measure-ments that should be made or, if the source does not yet exist, they will suggest the best way of predicting the likely noise levels and the optimum way of planning against nuisance hazard when the source is in existence.

Of course, identifying the source and measuring or predicting the levels of noise from the source, does not determine whether there is a reasonable case for noise control. This depends upon existing legislation, codes of practice or other agreed *standards*. These standards will relate to levels of noise that should not be exceeded or to the extent to which noise from an offending source may exceed a limit or to a specification of performance. The latter type of standard, for example, is invoked in Part G of the British Building Regulations (1965 amended 1972), which relates to the sound insulation of walls or floors dividing adjacent dwellings. There appears to be no similar US equivalent.

The standards that specify noise levels will do so in terms of noise units that depend upon the type of noise source being controlled. The fact that there are so many different noise units often leads to confusion and complexity. In the USA, a single unified noise unit, the day-night noise level (L_{DN}), is being advocated by the Environmental Protection Agency (EPA) to enable goals to be created for all environmental noise levels. However goals, like many standards, are not enforceable. The EPA further suggests that diversity of noise units should be retained in order to lay down enforceable emission standards on individual noise sources. In the UK another unifying noise unit L_{NP}, Noise Pollution Level, has been suggested but not accepted generally in legislation. So although the search goes on for a simple single sound measuring system, the reality today and for the immediate future means getting to grips with individual noise units for different types of source, such as traffic, aircraft and factories.

Even after appropriate units, standards, and methods of control have been taken into account in a noise problem, there are still many other factors to consider. For example, a solution may not be possible or it may cost too much.

It is not possible to solve every noise problem, so we cannot hope in this book to provide solutions in every case. Even those solutions we do suggest will involve the collection of more information in order to reach a successful outcome. For this reason the box labelled 'Ask for Advice' in Figure 1.1 is an important one. A large number of Universities and Polytechnics have departments and individuals who can offer advice. In the UK the Building Research Establishment has an advisory service, as does the National Physical Laboratory. In addition, there are many independent consultants in the field. The British Association of Noise Consultants (ANC) will provide a list of its members as will the American Institute of Noise Control Engineers (INCE) and its international partner, International INCE, to which many national associations belong. Most large metropolitan areas have a group within the environmental health or similar department who can be consulted on noise problems. In the United Kingdom the name and address of the local Environmental Health Officer can be found in the "Municipal Year Book" or from the local Citizens' Advice Bureau. Similar organisations can be found in other countries.

1.3 WHO CONTROLS NOISE?

The most important body of people who are or should be involved in noise control are the manufacturers of noise-producing devices, since in their hands lies the most effective way of controlling noise – at the source. However, we live in a society where even the most enlightened manufacturers need an incentive to invest in the extensive research, development, design and tooling that might be required to reduce noise emissions from their products. Such incentives are provided, in essence, by legislation enforced either centrally or locally. To appreciate the number of people and organisations involved in legislation for noise control and the ways in which they influence this legislation requires a look at the history of government concern with the problem of noise.

5

1.4 GOVERNMENT INTEREST IN NOISE PROBLEMS

Central government has been involved in matters of noise control in the United Kingdom since 1934 when the government of the day set up a committee to investigate the problem of traffic noise. This is a relatively short period of involvement compared with government concern over some other aspects of public health. For example, in the UK the awareness of the problems and investigation of the remedies in the area of air pollution culminated in the UK Clean Air Act of 1956, and since that time there has been a steady improvement in the quality of the air in large UK towns and cities. However, we are only just beginning to take the necessary legislative action against noise and as a result the improvement, if any, will not be noticed for some years. Indeed, concern at increasing environmental noise pollution is part of the growing awareness of problems of amenity rather than of health or death risk. As with other legislation for maintenance of public health and welfare, where there are no obvious short-term action benefits, it is possible to identify cycles in the formulation of legal powers which stem from informed awareness of the 'scientific community', through public outcry, government investigations, permissive then compulsory legislation. We look at these processes in more detail in Chapter 11.

The most important development after the institution of the 1934 Committee on Transport Noise was the commissioning of a survey of noise in homes in 1948 for the Building Research Station. This scientific advisory body, now part of the Building Research Establishment, has conducted a programme of research into many noise problems since that date. In 1960 the Minister of Science appointed a committee on the problem of noise under the chairmanship of Sir Alan Wilson (the Wilson Committee) "to examine the nature, sources, and effects of the problem of noise, and to advise what further measures can be taken to mitigate it". Also in 1960 the Noise Abatement Act received the Royal Assent; this provided a procedure for abatement of noise as a statutory nuisance by reference to the Public Health Act of 1936. In 1961 the Minister of Transport, Mr Ernest Marples, set up a study group led by Sir Colin Buchanan to study the long-term problems of traffic in towns. Their report, "Traffic in Towns", (see Bibliography) was published in July 1963 and made reference to the problem of traffic noise as a major factor in the deterioration of the environment. The group suggested that the long-term remedy must be by town planning, encompassing at one extreme a diversion of heavy traffic flows from residential areas and at the other extreme the detailed layout of buildings. Also in 1963 the Wilson Committee made its report, including a large number of recommendations about traffic noise and sound insulation. The Greater London Council was also taking an active interest in traffic noise with a policy statement in 1966 and the publication of a design bulletin on traffic noise in 1970 (see Bibliography). The successor to the Wilson Committee is the Noise Advisory Council, which is chaired at ministerial level and has issued various recommendations from 1970 on surface and air traffic noise, industrial noise, and acceptable levels of noise in residential areas. The Urban Motorways Committee, which was set up in July 1969 to examine the present position and to recommend what changes should be made to enable major new roads to be related better to their surroundings, reported in July 1972 recommending a limit to motorway noise at the facade of the nearest building.

This suggestion was backed up by the Noise Insulation Regulations of the UK Land Compensation Act of 1973. The latest legislation concerning noise control is to be found in the UK Control of Pollution Act of 1974, and this includes the powers given to local authorities to create noise abatement zones, another suggestion of the Noise Advisory Council.

Throughout the development of this legislation, as well as having scientific advice, the government of the day has been subject to various pressure groups including amenity organisations, the best known of which is probably the Noise Abatement Society, and the industrial lobbies including the Road Haulage Federation, now the National Road Haulage Association, and many others. There is no doubt that future projects, such as new or longer runways for airports and major urban roads, are certain to be the subject of objections on grounds of noise. Opposition will no longer be confined to protesting against the demolition of houses and the destruction of the character of local communities. Noise calculations can be challenged, providing a fruitful source of argument. The anti-road lobby is active in many countries and local groups have joined hands to wage war against proposed airport developments. There are some signs that the trade union movement will have a greater involvement in matters of working conditions, particularly those related to hearing damage risk. Planners, engineers, architects and other members of design teams, the economists and the decision makers at political level are beginning to take notice of these forces.

2. Noise and sound, noise units and their measurement

2.1 NOISE

Noise is defined as 'unwanted sound', which means that human beings, the recipients of the sound, are the ultimate judges of what is a noisy sound and what is not. Unfortunately for the purposes of the laws which regulate human conduct, it is not sufficient to leave the definition of noise like this since if, for example, a law was passed making it illegal merely to make an unwanted sound any person could at the drop of a hat render any sound-producing activity illegal.

Therefore the law must set limits and these must be specified in objective terms, eg readings on a meter rather than by complaints from (possibly eccentric) humans. However, the type of meter readings and the limiting levels set should be governed by the characteristics of sounds and by experiments or surveys designed to discover, as accurately as possible, the tolerance limits of human beings. In order to see how this is done we shall in the next few sections take a look at the nature of sound, its characteristics, and at how these can be related to human response.

2.2 SOUND AND ITS CHARACTERISTICS

Sound is caused by any disturbance in the air. When a stone is thrown into a pond the ripples from the point of impact spread out circularly. In an analogous way spherical pressure waves are set up in air moving away from a point of disturbance (Figure 2.1). Sound travels in air at a speed of approximately 350m per second. When the pressure waves meet the human ear they cause a sensation which we also call sound – although strictly the term applies only to the physical phenomenon. The human ear is a very sensitive measuring instrument for sound and can detect pressure fluctuations of the order of one ten thousand millionth part of an atmosphere. However, the ear also has a very large dynamic range that can cope with pressure fluctuations ten million times greater than this without immediate danger.

The general properties of sound usually have two names. This is because we have one name for the physical property of the sound in the air and another name for the subjective response of the human being to the sound. For example, the frequency of the air waves in Hertz or cycles per second is a measure of the rate of variation of air pressure at any particular spot. Our response to frequency we call pitch. The intensity of a sound wave is the

Figure 2.1 Spreading of sound waves in reflection-free conditions. A balloon, dilated and deflated by the action of a pump, represents the vibrating source. A sound wave radiates from the source as successive compressions and rarefactions of the surrounding air

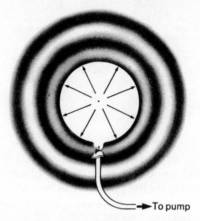

To pump

energy it carries per unit area. Our subjective response to intensity we call loudness.

If a sound consists of a number of separate waves with some form of harmonic relationship between them we may call the sensation music. A chaotic sound, such as that coming from a boiling kettle or a waterfall, produces a random signal which is technically called 'white' noise. It could be called noise in the subjective sense also. In this book the word 'noise' will be used to describe both chaotic sounds and musical/regular sounds whenever they are undesired.

2.3 SOUND INTENSITY

As the sound pressure waves travel through the air they also carry energy with them. The rate of flow of the energy is called the power of the sound. This power per unit area is very small, varying from 10^{-12} watts per square metre (that is, one million millionth of a watt) in the quietest sound, up to some thousand watts per square metre in the loudest sound. At normal speech levels the amount of sound power falling on a square metre is 10^{-5} watts. When you consider that an electric fire radiates energy at the rate of two or three thousand watts, then you can see that sound is not a very efficient means of transferring energy.

2.4 TYPES OF SOUND FIELD

Sound can exist as free travelling waves or as standing waves confined to a finite region. For example, when a sound travels between two parallel reflecting walls, it can be shown that the two travelling waves passing in opposite directions between the walls combine to give a standing wave. There is no net flow of energy through space, but there is a pressure fluctuation which varies in amplitude from point to point between the walls. The ear can hear this sound field. Such sound fields are called 'reactive' in analogy to the electrical reactive impedance which has similar properties of energy storing and current limitation without absorbing energy (by degrading it into heat). Other forms of reactive sound field occur near small sources and along boundary surfaces, for example the outside skin of an aircraft.

2.5 THE EAR AND HEARING

Figure 2.2 shows the human ear. It is necessary to have an understanding of how the individual parts of the ear behave so that when we come to discuss hearing loss and deafness we can appreciate what is happening.

Sound first enters the ear through the outer canal. This is a small tube about 25mm long and varying in diameter. Little physical damage can be done to this tube except that it does tend to collect low grade bacteria and once a colony of these is established it is very hard to shift. The sound then falls on the eardrum, which is not flat but is a cone with an included angle of about 120°; the apex is on the inside of the eardrum. The eardrum vibrates under the influence of the incident sound wave. This vibration is then connected through a series of bones which reduce the amplitude but increase the force upon inner 'windows' which, in turn, transmit the waves to liquid borne pressure waves within the circular canal called the cochlea. The pressure fluctuations within the liquid in

Figure 2.2   The ear mechanism. The round window allows the pressure wave in the cochlea fluid to dissipate. The semicirclar canals sense the motion of the head. The eustachian tube connects the space behind the eardrum with the mouth cavity

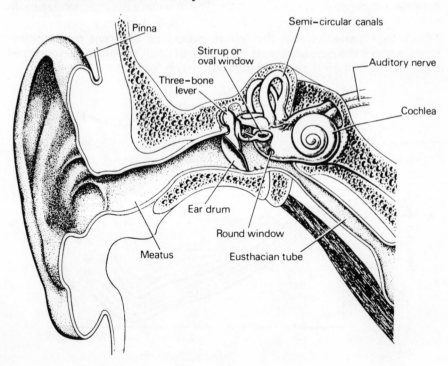

the cochlea then excite small nerve cells which are called hair cells. Each of these hair cells has its own individual nerve and these nerves are connected together into the auditory nerve which then goes off to the appropriate brain centre.

It is important to remember that the human brain is a component of this system. The brain works as an on-line computer interpreting the signals it receives from all the hair cells in the cochlea. There is a lot of parallel information coming in along the various nerves and it is a mistake to think for example that an individual nerve cell corresponds to a particular frequency. In fact, groups of nerve cells together tend to be responsible for human aural response at particular frequencies. Because of the complicated nature of this system it can be seen that the human ear is likely to react in different ways at different frequences of incoming sound. In fact, the ear tends to be most sensitive at about 3000 Hz because of a resonance in the outer ear canal. The sensitivity of the ear falls off rapidly at high and low frequencies and a graph of the relative sensitivity of the ear is shown in Figure 2.3. It can be seen that complicated effects set in when the intensity of the sound varies, so that whereas at low intensity the frequency response of the ear varies quite markedly, at high intensities (about 90 to 100 dB) the subjective response of the human ear is, for all intents and purposes, flat (that is, independent of frequency). (Decibels are explained in Section 2.8).

2.6 DAMAGE TO THE HEARING MECHANISM

The ear can be damaged in a number of ways. Physical poking of objects into the outer canal can rupture the eardrum. This is usually not completely fatal to

Figure 2.3 Sensitivity of the ear to sound of different frequencies. Curves show the percentage of 1200 ears of healthy young people aged between 18 and 25 that could just hear each test tone

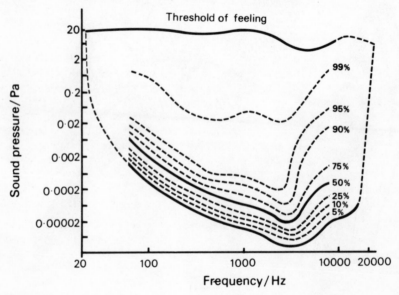

the hearing mechanism and eardrums can be repaired. A very large shock wave can sometimes physically break the bones of the middle ear which transmit the sound across from the air to the lymphatic liquid in the cochlea and when this happens instantaneous deafness can occur, but again this can often be repaired by skilful surgery and prostheses. Once the sound waves have entered the cochlea they are then picked up by the hair cells and it is these hair cells that are damaged by prolonged exposure to high intensity sounds. It is best to think of these hair cells as slow blow fuses. They can withstand a short period of high intensity sound but then start to deteriorate and are broken. It is an unfortunate fact that in general nerve cells within the human body do not regenerate themselves. So that once a hair cell has been damaged it is totally lost.

Over a period of a lifetime progressive deafness occurs as these cells die off. This is called presbycusis. Many people think that presbycusis is a natural phenomenon, but others think that it is partly a result of the high noise levels that we have in our society. In fact, many audiologists (scientists concerned with the ear and hearing) have adopted another term, sociocusis, to represent the loss of hearing caused by general, non-occupational, noise exposure. It seems probable that people who live in urban areas have less sensitive hearing than those who live in remote rural areas and who experience little or no industrial noise. Indeed, the self-inflicted noises of many of our leisure pursuits – such as discotheques – are making sociocusis an increasing reality.

It is now well understood that excessive noise can cause damage to the ear: in fact, any continuous noise level above about 85 dB(A) will cause some damage. This damage is slow, insidious and progressive and is not usually noticed by the recipients until it is far too late. A lot of people who have been working in high noise level industries all their lives are now quite deaf in their old age. This is a subject which is becoming increasingly of interest in many countries because it has now been established that employers can be held liable for deafness caused to their employees merely by noise alone.

2.7 LOUDNESS AND ANNOYANCE

There is no doubt that the annoyance caused by sound is most closely related to the loudness of that sound. Such effects as masking and other sensual stimuli tend to distract the attention from a particular sound but when that sound becomes so loud that it intrudes upon the sphere of activity of the noise-exposed individual then he notices it. Quite often he chooses to ignore it. This can be for many reasons: he knows what it is, it is something which he has control over, or it is something that he tolerates because he happens to be that sort of individual. In a minority (about 20%) of the population, high sensitivity to noise is shown. A larger group of the population (about 60%) will tolerate quite loud sounds and are not particularly disturbed and there is a small minority who even though they are not deaf, will hardly ever complain about noise.

Thus many standards of measuring noise and annoyance relate the annoyance basically to the loudness of the sound. Corrections are sometimes made for other factors such as pitch, duration, whether or not the sound is impulsive, and whether it goes on during the day or the night.

2.8 MEASUREMENT OF NOISE

The measurement of noise is a subject complicated by the difficulty of trying to reproduce on an electronic instrument a meter reading that bears some relationship to the thing that really counts – the subjective response of human beings.

The units of measurement of sound and the instruments designed to carry out the measurement must fulfil one or more of the following conditions:

1. Provide objective information about the physical phenomena that are taking place.
2. Provide a measure of the acceptability of a sound.
3. Provide a scale upon which legal limits can be fixed.
4. Measure the seriousness of the offence of exceeding a limit.
5. Give information about damage risk and health hazards.
6. Give a means of setting compensation levels and insulation requirements.

In view of these diverse requirements it is not surprising that many different units of sound have sprung up. If we add to these the fact that sound may have musical qualities and information content as well as mere loudness and noisiness, the design of a single sound measuring device capable of giving objective information about sound is seen to be nearly impossible. However the attempts to achieve the above aim usually start with the decibel.

The decibel
The decibel is the basic unit of sound and it is the unit designed to fulfil the scientific requirement of giving information about the physical amplitude of sound. It can be and is applied to noise assessment, but given that its principal purpose is scientific rather than social we must use it with care.

In air, sound consists of pressure fluctuations about the mean atmospheric pressure. To be audible these fluctuations must have frequency (or frequency components) between 20 Hz and 20 000 Hz. If the amplitude of pressure fluctuations is P then the *sound level* in decibels is given by:

$$L = 10 \log_{10} \left[\frac{P}{P_0} \right]^2 \text{dB}$$

where $P_0 = 2 \times 10^{-5}$ N/m² (a reference pressure)

The dB is thus based on a ratio of pressures and so is dimensionless.

The reasons for the logarithmic nature of the decibel are:

1. The amplitude of audible pressure waves varies from 2×10^{-5} N/m² to above 10^3 N/m² which is a vast range.

    By taking the logarithm of these extremes we reduce the scale to roughly 0 to 15 Bels and by multiplying by 10 we bring it to a practical range of 0 to 150 dB (decibels)

2. It has been found that, by and large, people judge that a sound which has increased in level by 10 dB is roughly doubled in 'subjective loudness'.

It is, however, a mistake to take this too literally and then assume that an increase of 40 dB represents a loudness increase of 16 times. The decibel is useful for handling the large variations in pressure amplitude that cause small variations in subjective response.

Note that we measure sound *level* in dB. The word 'level' whenever you see it should remind you that you are dealing with a logarithmic measurement. The unit called the decibel, by the way, is used whether we are talking about sound pressure level or sound power.

$$L = 10 \log \left[\frac{P}{P_o} \right]^2 = 10 \log \frac{W}{W_o}$$

(where W_o is a reference power 10^{-12} watts per square metre)

Addition of decibels
The logarithmic nature of the dB unit causes a lot of difficulty when variations in sound level are considered, but the correct use and understanding of the decibel is essential for any person endeavouring to control noise.

An important aspect of the scale is that whenever sound power is doubled – for example, by adding two identical sources together, the resulting sound level is not doubled. The fact that a single motorcycle will create 100 dB at a kerbside does not mean that two motorcycles will create 200 dB at the same point. The result of adding two identical sound levels together means an increase of only 3 dB in the sound level from one source. So if one motorcycle produces 100 dB, two together will produce 103 dB, four will produce 106 dB and so on. So you can see that an increase of 10 dB represents a tremendous increase in sound power or pressure. But subjectively a 3 dB change in the sound level is just noticeable, and 5 – 10 dB change is required to achieve substantial change in subjective appreciation of noise level.

However, we must take care in deciding how to add sounds together: we consider this in more detail in the next subsection.

Incoherent sounds
Coherence is a property that relates one sound wave to another. If any sort of physical relationship between two sounds exists then they may be coherent. Two pure tones of the same frequency are always coherent and complex tones may be. White noise from two sources is always incoherent in that no regular relationship exists between the two chaotic signals. When two incoherent sounds are mixed the sum of their independent power levels is added. The safest way to add two sounds is to assume that they are incoherent and to add their powers. Power, whether in the form of propagating waves or reactive sound fields, is related to the sound level by a logarithmic relationship of the form:

$$L = 10 \log_{10} (W/W_o)$$

where
W is the power, L the sound power level and W_o an arbitrary power reference, usually 10^{-12} watts per square metre.

When two sounds are added, we must thus determine the power levels of each, add these and then determine the power level of the resultant field. For example, if

$$L_1 = 10 \log W_1/W_o$$
$$L_2 = 10 \log W_2/W_o$$

Inverting these, we have

$$W_1 = W_o \, 10^{(L_1/10)}$$
$$W_2 = W_o \, 10^{(L_2/10)}$$
$$W_3 = W_1 + W_2 = W_o(10^{L_1/10} + 10^{L_2/10})$$

So

$$L_3 = 10 \log (W_3/W_o) = 10 \log (10^{L_1/10} + 10^{L_2/10})$$

Notice that W_o has vanished from the final result.

The reader should try putting a few numbers to the formula to familiarize himself with the results.

Example: add a sound of level 66 dB to one of 79 dB

$$L_3 = 10 \log (3.98 \times 10^6 + 7.94 \times 10^7)$$
$$L_3 = 10 \log (8.341 \times 10^7)$$
$$L_3 = 79.2 \text{ dB}$$

This is a negligible change on the louder sound level but an important result from the point of view of noise control.

Coherent sounds
Here the subject is much more complicated and almost any result from silence to plus 6 dB is possible when equal coherent sounds are added. For a discussion of this topic see Section 8.6.

2.9 OTHER NOISE UNITS

All this convenient mathematics fades into uselessness when the real problem of assessing the noisiness of sounds is tackled. Sound contains pitch and character, and the human brain is well able to pick out sounds of a particular quality against background noise levels that are 20 dB or 30 dB higher. So it is quite easy for noises that do not register on a sound level meter at all to be quite audible.

Persevering with the problem of constructing a unit to measure human response to noise, we come to the dB(A). This is a method of allowing for the fact that over the usual range of sound pressures, the human ear has a particular sensitivity which depends upon the pitch (or frequency) of the sound that it hears. The ear is most sensitive at 3 kHz and has much lower sensitivity at high and low frequencies. So a means has been devised where, by placing a frequency filter in a sound level meter, the meter is made to respond with a frequency sensitivity similar to that of the human ear (see Figure 2.3).

Unfortunately, the human ear's frequency sensitivity varies with amplitude (loudness) and is considerably flatter at higher amplitude. Other scales (dB(B) and dB(C)) were originally designed to meet this fact but these scales have fallen into disuse, leaving the dB(A) as the common scale for use with broad-band noises (that is, noises containing most frequencies in the audio-frequency range).

Units for measuring intermittent noises
When a noise is not constant but varies with time it is often important to have information about the time history of the noise so as to give information about the dose of noise received. This is important in a number of contexts. In the workplace, we are interested in the equivalent continuous noise level received by an employee over his working day – L_{EQ}. In the area around an airport, we want a measurement of the noise received by people living in the area. For this purpose both the noise level from each aircraft and the number of aircraft is important. This is discussed further in Chapter 4.

People living near to a busy road suffer noise all the time but the maximum level is thought to be the important factor in deciding their annoyance. See Section 4.4.

Noise dose
It is known that exposure to loud sound causes progressive irreversible hearing loss and it is also known that hearing deteriorates with age (pre-sbycusis). In order to assess the damaging power of a day of exposure to noise levels, the idea of a noise dose has been evolved. The idea here is that a certain amount of sound energy can be tolerated in a working day, but that above this amount damage occurs. The damage risk level is a subject of much controversy. In some countries it is set at 85 dB(A) for 8 hours, but in relevant legislation in the UK and the USA a higher dose is prescribed: 90 dB(A) for 8 hours.

For higher levels of sound, the same dose of sound energy can be received over a shorter time (see Table 2.1).

Table 2.1   Exposure levels and time limit (UK)

Level	Dose time limit
90	8 hours
93	4 hours
100	48 minutes
110	4.8 minutes
120	28.8 second
130	2.88 seconds

On an equal energy basis, an increase of 3 dB in the exposure level may be permitted for each halving of the duration of exposure. However, increases in level cannot be tolerated on this basis indefinitely. When the level is increased over a shorter time above 130 dB, hearing damage can occur instantaneously.

Combinations of levels are integrated to give an equivalent level, which, if experienced continuously during the day, would give the same total dose as that actually received. Further details of this calculation are to be found in

Chapter 9. The calculation and measurement procedures are straightforward where the exposure level is steady or varies by less than ±3 dB(A). However, if the level fluctuates by more than ±3 dB(A), it is necessary to measure the equivalent continuous level L_{EQ} (pronounced L-E-Q). L_{EQ} is the constant level that would produce the same amount of energy at the measuring point as the actual fluctuating level during the measuring period.

Noise dose meters based on the formula used in 'Code of Practice for reducing the Exposure of Employed Persons to Noise' are available to check the level of noise received by persons thought to be at risk of noise induced hearing loss. These meters give L_{EQ} directly and are reasonably accurate, based as they are on the same technology which has given us the hand calculator which they closely resemble.

This is the principle used in the UK and in many other countries. However, in the USA, Belgium, Italy and Canada an increase of 5 dB(A) in the level is permitted for each halving of the exposure time, rather than 3 dB(A). This 5 dB(A) increment is intended to allow for the expected intermittency of exposure even to steady levels. For example, individuals exposed to noise will have natural breaks from the noisy environment. Nevertheless, the consequence is that the standards based on an equal energy basis are stricter.

Percentile levels: L_{10} (L_{50}, L_{90})

When measurements of road traffic noise are made we are dealing with a noise source that is strongly time dependent. In the UK the measure that is used is not an average or an accumulated dose but an approximation to the maximum level, defined as the level in dB(A) exceeded for 10 per cent of the time, L_{10} (pronounced L – TEN), during every hour over a period of 18 hours from 6 a.m. to midnight on a typical working day. The average of the 18 L_{10}s found during this process is called the '18 hour L_{10}'. This noise measure is used for the purposes of the Noise Insulation Regulations of the Land Compensation Act (1973 Amended 1975). This legislation is reviewed in Chapter 4, together with details of measuring L_{10}.

Other statistical measures of fluctuating noise levels can be specified in a similar manner. For example, L_{90} (pronounced L – NINETY) is the level exceeded for 90 per cent of the time. L_{90} is often referred to as the background level of noise. One of the accepted schemes for predicting traffic noise in the USA predicts L_{50}, the level exceeded for 50 per cent of the time.

L_{DN}

In the USA the day-night sound level has been selected by the Environmental Protection Agency as the preferred unit for evaluating interference of noise with human activities, for example, speech communication, sleep and other activities where annoyance might be caused.

L_{DN} is essentially the A-weighted 24 hour L_{EQ} with a 10 dB(A) penalty applied to the period from 2200 to 0700 hours.

2.10 CONCLUSION

This concludes a brief introduction to a complex field, which we believe contains the basic points needed for an understanding of noise units. Care must be taken when using instruments and units, especially if a hard fight develops about a particular noise problem. Whether or not he is in the right, the amateur and even the professional can be made to look foolish in court or at enquiries. Study all the available literature carefully if you progress with a noise problem, particularly the manufacturer's handbook when using instruments, and always consult the relevant legislation to make quite sure that you are considering the appropriate standards, measurements and units or indices. Any point that you do not completely understand must be researched carefully, so that if you are put to the test you will have an answer – the correct answer.

3. Principles of noise control

3.1 INTRODUCTION

It is perhaps unfortunate that once noise has been established in the air, there is almost no way in which it can be controlled by means of a power-driven device. We have devices to vary temperature, lighting levels and the condition of the air and so these aspects of human comfort are well within our control. This is not so with noise; there is no electrical 'noise level controller' that can be bought and installed in homes and work places. Worse than this, sound is in fact produced by many of the devices that are used to improve our environment, so that the world abounds with many sources of sound but with no ready means to control noise levels.

How, then, is noise controlled? It is controlled by passive means, means that do not consume energy and which therefore cannot be controlled by a switch. Noise control can be achieved mainly by planning and forethought. The passive barriers, isolators, distances and absorbers which lead to an acceptable noise level are usually permanent features of our environment and as such they will not be easy to adapt or vary, and so must be adequate when installed, and allow for possible increases in future noise levels.

It must be remembered, also, that noise control is often a compromise. Silence can only be achieved in two ways: either the noise source is switched off, thus depriving someone of a facility, or else the noise sufferer must move away until he can no longer hear the sound. If neither of these two solutions is acceptable, the 'noise expert', when faced with the demand "Stop that noise", must admit that it cannot be done, but that it may be possible to reduce the noise to an acceptable level. The concept of an 'acceptable level' is a difficult thing, and often involves the setting of standards by an arbitrating body such as government and courts of law. Decisions will have to be made involving a trade off between utility and noise. An example here is the person who joins his local aircraft noise abatement society and then complains when he cannot fly to anywhere at the drop of a hat from his local airport. He cannot have air transportation without aircraft noise, and must accept some compromise noise level while directing attention to the most important element in any noise problem – the noise source. In this case control of the noise source is linked with the design of aircraft propulsion systems, or even the design of less noisy means of rapid transportation.

This then is how the problem will be considered: noise control at the source

21

of the noise, along the path that the noise takes and at the point of reception. Over all these aspects is the necessity to plan well and to educate people and organisations to consider how avoidable noise can be eliminated and how unavoidable noise can be reduced.

3.2 GENERAL NOISE CONTROL FACTORS

The energy contained in a sound wave
When work is carried out, it can only be produced by means of a heat engine or an electric motor. There is no other source of power, and the ultimate end of all work is waste heat. One step above waste heat comes sound energy: a minute quantity of the work done by most engines is not directly converted into useful work or into waste heat but is radiated as sound. It is the smallness of the quantity of energy needed to produce a sound that is a major factor in the difficulty of noise reduction. A machine produces a mere 10^{-6} to 10^{-9} of its total output as noise, but this minute fraction of sound can represent a noise level of a shattering 100dB(A) or more, such is the sensitivity of our ears. Not only that, but if the noise has a particular characteristic, such as a tone, or an impulsive character, or an intermittency, or if it carries information, the brain will rate the annoyance value of the noise at an even higher level than a straightforward measure of the sound level would indicate. Thus in trying to control sound we are chasing after a minute fraction of the energy involved in the work-producing process.

It is therefore difficult at the outset to redesign an existing device so that it is inherently less noisy. On the other hand, if a device has been badly made it can easily put out a much larger fraction of its power output as noise. In this case, noise control is much easier. However, there are not many things that are built to be unnecessarily loud, unless that was the intention of the designer and the user.

Table 3.1 Loudness, some sounds

Noise	dB(A)
Large jet airliner	140
Large piston-engined airliner	130
Riveting on steel plate	130
Train on steel bridge	110
Weaving shed	100
Pneumatic road drill at 30m distance	90
Heavy road traffic at kerbside	85
Dining Room	80
Male speech at 1m distance	75
Typing office with acoustic ceiling	75
Light road traffic at kerbside	55

It is, of course, pointless to class devices which are intended to produce noise, such as the loudspeaker and the football rattle, as controllable noise sources – their designers and operators would usually wish them to be even louder if possible.

Outdoor sources

First we must consider the types of noise source that must be controlled. Amongst outdoor noises, road traffic noise is the most widespread and usually the loudest. In recent years, this problem has become so intense that special measures, such as the Land Compensation Act, have been taken by the government to limit the growth of this noise and maybe even to reduce it.

A more localised form of noise, but one which produces levels far greater than that produced by road traffic, occurs around airports. Peak levels well over 110dB(A) can occur miles from the airport and whole communities can suffer continued exposure to extreme amounts of noise.

A third important source of noise is that due to construction. A construction site, even though it is a temporary thing, can produce a great deal of annoyance during operation which can last up to three or four years. This annoyance is exaggerated by the resentment that persons suffering from the noise feel about the change in their environment caused by the construction work and this tends to exacerbate the situation.

Noise from factories is a widespread source of complaint but reasonable procedures exist for dealing with this phenomenon. Finally, there is domestic noise from one's immediate neighbours which very often has no general solution, but has to be dealt with as individual cases.

Indoor sources

Indoor noises principally concern the control of noise within a factory and other places of work, and are receiving increased attention from the government. It is not yet illegal to produce any particular noise level within a factory except within woodworking areas but a Code of Practice exists which lays down guidelines which are likely to become statutory, and trade unions are having

Figure 3.1   Positions of measurement for small machines

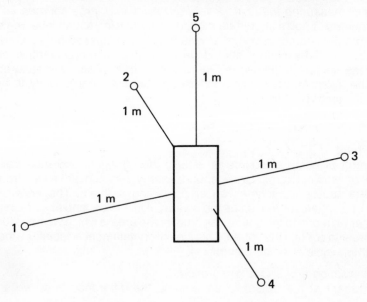

Figure 3.2 Positions of measurement for machine tools and other stationary plant. Measurements should also be taken at any other positions likely to be occupied by a person for a significant length of time

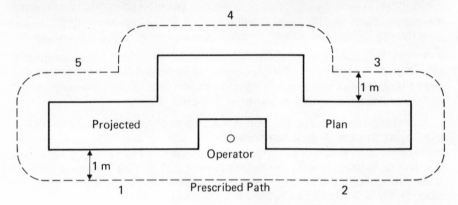

increased success in damage actions for noise-induced hearing loss. The control of indoor noise is therefore important. Requirements of privacy and intelligibility for conversations in offices mean that the quieter noises in offices must also be controlled. At the bottom end of the loudness scale, the stringent requirements for quietness that exist for dwelling houses mean that noise control of service systems requires attention.

The approach to noise control
We will now continue to consider these various noises and the methods of controlling them. Noise control methods naturally break down into three parts: noise control at source, noise control along the path that the noise must take to reach the receiver, and noise control at the receiver, which includes his immediate acoustic environment.

Method of measuring the noise level from small and large sources
Small sources of noise should be measured by placing sound level meters 1 metre from the source, at points around the source, spaced out as shown in Figure 3.1. A large source should be measured at representative points around the source as shown in Figure 3.2. In both cases, a measurement should be taken at any position where an operator's ear is likely to be for substantial lengths of time.

3.3 NOISE CONTROL AT SOURCE

Containment
The control of noise at source is usually the cheapest and most reliable method of noise control. Once sound energy has moved away from its immediate source it becomes an all-pervading problem. The area of the surface of a sphere increases as the square of its radius, and this means that the nearer we move to the source the smaller the area is which we have to deal with in order to contain the sound. This idea of containment or isolation, then, is the main principle of noise control.

Sound insulation and sound absorption
It is important at the outset to understand the difference between sound

24

insulation and sound absorption: these two different but complementary means of noise control are often confused. Sound insulation concerns the reduction of sound as it passes through a wall or barrier, or, in the case of vibration, through isolators or discontinuities. Here the major contribution to the reduction in sound level is the fact that by far the largest fraction of the sound energy is reflected back from the barrier. For example, if a wall has a transmission loss of 30 dB, this means that only a thousandth part of the energy falling on the wall is transmitted but up to 999/1000 parts may be reflected back. Sound absorption, on the other hand, occurs when a sound is partially reflected from a wall. Sound is absorbed in many ways at a boundary and the reflected sound is reduced because of these absorbing processes. With good sound absorbers 90 per cent of the incident sound energy can be absorbed. Perhaps the most significant difference between insulation and absorption is the fact that absorption involves a degradation of sound energy into heat whereas insulation can be achieved without any such degradation at all.

Enclosures

General principles. If it is desired to reduce the noise from a particular source, this can often be done by means of an enclosure. However, the mere surrounding of the source by massive partitions is not adequate, since the volume enclosed with the source becomes a reservoir for noise energy. What happens then is that sound levels inside the enclosure increase until the rate of loss of energy within the enclosure is equal to the rate of production of sound energy from the source. If the only way in which sound can escape from the enclosure is through the wall, it is immediately obvious that the sound energy will build up until the rate of radiation of sound from the enclosure is equal to the rate of production of sound from the source and the situation will be as if the enclosure did not exist at all (except for directionality effects). It is therefore necessary to include within the enclosure an alternative means by which the sound energy can be converted into heat. This is done by including sound-absorbing material within the enclosure and when this is done most of the sound energy is lost into the sound absorber and the insulation becomes effective.

Access problems. A critical problem with barriers is that of permitting access to the machinery or noise source that is enclosed. Access can be required for a number of reasons: for monitoring, adjustment of controls, the supply of raw materials and the removal of finished products and also for the removal of waste heat.

With a little ingenuity most of the control problems can be overcome by providing a small window and access hatch, preferably double glazed, in the barrier. Gauges can be read without opening the barrier and controls that only have to be adjusted occasionally can be left within the barrier. Controls which have to be continually adjusted require special treatment and this can often mean re-designing the machine so that the barrier becomes an integral part of the whole device. This has to be dealt with at the time of specification of the machine, which we will discuss below, but in the case of an existing machine it is often possible to provide extensions to controls, or in the case of electrical equipment to remove switches and rheostats to a remote position.

For continuous process machinery, the through put of materials is a more demanding condition. It is often necessary to locate the immediate operator of the machine inside the barrier area, in which case the operator must either bear a noise greater than that recommended, which involves the use of ear defenders more or less permanently, or if the operator is a member of a team carrying out the work which involves less noisy operations, the team can be cycled through the working day so that no individual operator exceeds the noise dose limit.

The final reason for providing access is the matter of cooling. Unfortunately, all materials which have good sound absorbing properties also tend to have good thermal insulating properties, which means that a temperature sensitive process cannot be enclosed in sound insulating barriers. However, it is possible to provide access for cooling air by means of chimney-like constructions. A duct, although it provides a direct air path to the surrounding area, does provide a certain amount of attenuation, particularly if the duct is lined with sound absorbing material.

Handbooks are available from duct manufacturers quoting the value of sound attenuation that can be expected from a given duct cross-section and length. These should be used to calculate the length of chimney required to provide adequate sound reduction as well as ventilation.

3.4 MACHINERY DESIGN AND SPECIFICATIONS

When ordering new machinery to be installed in a factory, it is now prudent to consider the noise that will be produced by the machinery so as to ensure an acceptable working environment for employees, about which legislation seems likely. The writing in of noise specifications would appear to be reasonably straightforward, but the main difficulty arises out of the interaction of the machine with the environment where it will be installed and the difficulty of measuring the noise produced by the machine in the environment where it is manufactured. It would be convenient merely to specify that the machinery when installed in the factory should produce a noise level not exceeding XdB(A). However, this places the supplier of the machine in a difficult situation. He does not know what acoustical environment exists within the factory, but he can measure the noise level from the machine within his own environment. The method employed is usually to specify that the noise level measured at a distance from the machine should not exceed X dB(A), and to compute the resulting noise field from a simple estimate of the acoustics of the room in which the machine is to be installed. In most cases, the noise level in the room will not exceed the level measured close to the machine. Exceptions occur when the machine is placed by itself in a small, highly damped chamber. Under these conditions, the direct field is dominant and the noise level must be calculated from inverse square law and directionality.

The design of machines for quietness is technical and specialised. A few general rules can be given here, but individual problems would have to be studied on their own merits with the aid of more technical information. The general principles are as follows:–

1. Impacting parts of machines should be enclosed.
2. All enclosures should contain sound-absorbing material suitable for the environment within the enclosure.

3. Manufacturing tolerances should be kept as small as possible; rotating machinery should be dynamically balanced.
4. All rotating or impacting machines should be based on anti-vibration mountings.
5. All rigid connections of the machine in the way of electricity, water, air, gas, etc, should have vibration decouplers around the machine and water ducts should contain pressure release chambers.
6. Internal combustion engines should be properly silenced.
7. Machines should not be sited in small reverberant chambers, but placed within absorbent enclosures sited in large work areas. The over-all layout of a building or factory should be designed so that noise critical areas (offices, show-rooms etc) are kept away from noisy areas by means of buffer non-noise critical areas, such as kitchens, canteens, lavatories, corridors, etc.

3.5 VIBRATION, ISOLATION AND DAMPING

Machines and other sources of impact noise (principally footsteps) can feed acoustic energy into the frame of a building in the form of vibration which can then run throughout the building and, by re-radiation from walls, cause noise problems at quite remote parts of the building. It is now known that under circumstances where a room has a resonant frequency at the frequency of some remote vibration source, such as a pump, the noise level in the room can become high due to tuned resonance effects. Adjacent rooms without tuned resonancies can be unaffected by the noise. Such anomalous forms of high sound level can also be caused by central heating systems where the large area of the radiators can act as efficient acoustic radiators and can couple the sound within the pipes to the room, producing a high noise level.

The principle cure for these problems is to provide discontinuities. If a building is constructed in a single steel frame, then the vibration energy can run through the frame virtually unhindered and produce a noise problem wherever high coupling or resonance effects can occur. Discontinuities can be introduced into the system by bolting the framework together with lossy pads between adjacent members, thus attenuating the vibration energy as it passes through the structure. In direct analogy with the airborne case, however, it is found to be most effective to cut off the vibration energy at source.

We must first distinguish between load-bearing connections from the vibration source and non-load-bearing connections. Obviously the weight of the machine must be supported somehow and this has to be done by providing a strong suspension for the mass associated with the machine. The principal problem here is that the degree of freedom introduced by such suspension leads to a resonant frequency. It is necessary to keep this resonant frequency down in the region where little vibration energy exists and to fit override stops to prevent occasional high vibration amplitude during, for example, run up on a machine when high low-frequency amplitudes will exist for a short period of time. Figure 3.3 shows the isolation that can be obtained from such suspension. Figure 3.4 shows the typical natural frequencies obtained with various common methods of suspension.

Figure 3.3 Vibration reduction (dB) by resilient machine mountings.

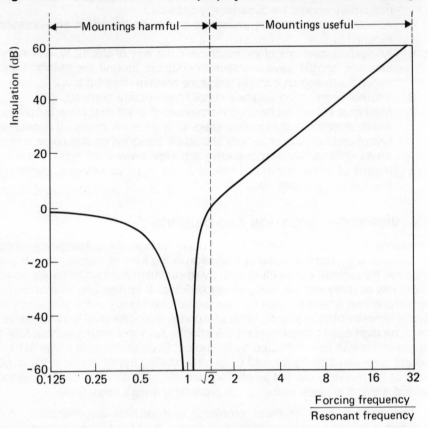

3.6 NOISE REDUCTION ALONG SOUND PATHS

Introduction

Once sound has left a source and become established in the surrounding medium, either air or the structure of the building, it must travel some distance before reaching the point at which the noise nuisance will occur. When this situation arises, we must consider steps that can be taken to reduce the transmission of noise as it travels down the path. Common paths for noise are airborne paths such as ducts and corridors and it is usual to include in this category walls that break up airborne sound paths, and not to consider these walls as independent vibration paths.

Alternatively, vibration energy can be transmitted through building structures directly and can arise from direct excitation from the source, or indirect excitation through a sound field produced in the room containing the source. Where a noise source is directly coupled to conducting paths such as pipes or air ducts, these paths can carry sound energy.

To control noise along the path, then, we need to consider how sound is transmitted through ducts, corridors, directly through walls and along pipes, and through the structure of buildings.

Transmission of sound along ducts

The manufacturers of fans and ducting systems include some who have paid great attention to the problems of noise. It is possible to contact them directly at the design stage of a building and they will quote sound levels and sound insulation values together with prices and delivery information for any reasonable noise requirement. Before talking to such people, however, it is advisable to have an understanding of what is a 'reasonable noise requirement' and what problems need to be considered. It is not reasonable to expect total silence from a system but it is reasonable to expect sound levels below ambient noise levels when the building is in use. The noise produced by the system should be graded to the existing noise climate of the building and will be related to the power of the driving fan (Figure 3.5).

A second point is the fact that ventilating systems form a link between the inside and the outside of a building and thus noise can be transmitted through the system and can be produced outside the building by the system itself. It is not usual for such noises to cause trouble except in the case of large industrial plants where the use of giant centrifugal fans can cause discrete frequency noises (at the frequency at which the blades of the fan pass the outlet), or in the case of discotheque noise in built-up areas. Conversely, when a ventilating system is installed in a noise-critical area such as near an airport the noise transmitted into the building from outside must be considered.

In all cases, therefore, the three factors to consider are: the noise produced by the system inside and outside the building, and the noise transmitted through the system.

Figure 3.4 Relation between static deflection and natural frequency

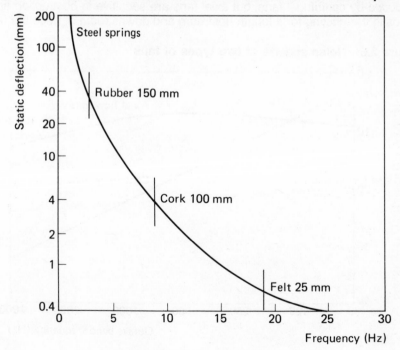

29

Figure 3.5   Noise level in a normal room as a function of fan power

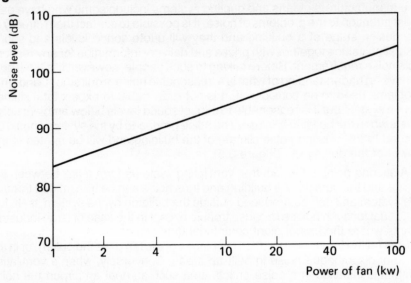

Noise from fans can be found from references to manufacturers' information or by empirical formulae derived by Beranek. Figure 3.6 shows a chart of expected sound power level. This is in the form of a "spectrum". A spectrum is a chart showing the variation of loudness as a function of frequency. The coupling of a fan to its associated duct does not greatly affect the noise produced by centrifugal fans, but axial fans are sensitive to obstruction and bends in the ducting for a length upstream and downstream of twice the fan

Figure 3.6   Noise spectra of two types of fans

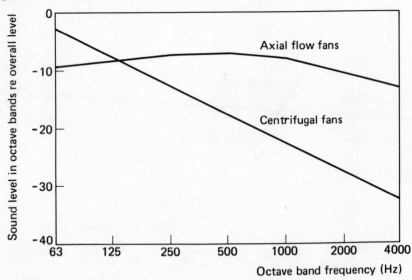

diameter. It is necessary to point out that sound radiation from a fan unfortunately travels equally well upstream and downstream. This is because the speed of sound (343 m{sec) is in excess of any air speed used in practical systems. It has been found that circular ducting is inherently more rigid than rectangular ducting and thus circular ducting transmits sound along the duct to cause a problem at the far end, whereas rectangular ducting tends to allow sound to pass through walls of the duct, possibly causing problems near the fan.

There are commercially available duct attenuating systems but it is necessary to remember that noise is transmitted down a duct in two ways. Firstly, it is transmitted along the air path within the duct. To counteract this, the provision of sound-absorbing material within the duct is necessary. The achievable attenuation depends upon the sound absorption coefficient of the lining (Figure 3.7). The absorption coefficient will depend upon frequency and upon the type of sound absorber employed (Figure 3.8). Secondly, noise is transmitted along the rigid walls of the duct. To attenuate this noise, it is necessary to incorporate a non-rigid bellows section to de-couple the duct from the vibration produced by the fan. Bends within the ducts will attenuate noise travelling along the duct but can cause noise in their own right if turbulence is generated at the bend. The provision of aerodynamic guide vanes within the duct can reduce such noise. At the end of a duct the sound must pass from the duct into the air space of the room. A useful effect occurs here: that of end reflection. This phenomenon occurs at the open end of an organ pipe and helps it to resonate. In the case of general noises travelling along the duct, it is possible that as much as 90 per cent of the sound can be reflected from the end. It is to overcome this effect that the mouths of wind instruments are belled, permitting the efficient coupling between sound within the tube and the surrounding air. The aim of the designer of ventilating systems is to achieve the opposite effect by providing low coupling to the room while avoiding localised noise sources which are often caused by turbulence produced within constricting grills. Much more detailed information on this extensive subject can be obtained by referring to literature produced by duct manufacturers.

Figure 3.7 Attenuation of a lined duct

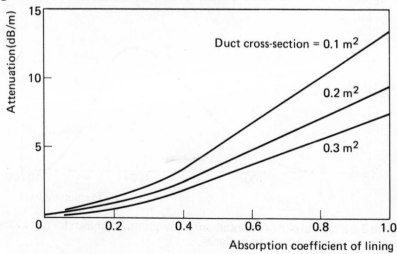

Figure 3.8 Sound absorber frequency characteristics

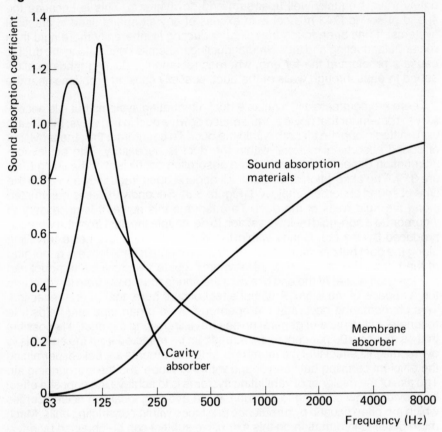

Figure 3.9 Frequency distribution of lined duct attenuation

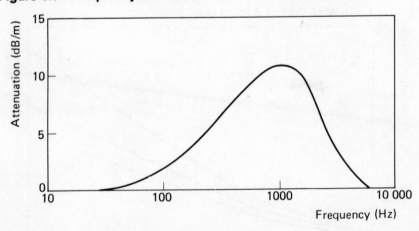

In Figure 3.9 a typical plot of attenuation vs. frequency is shown for a duct lined with porous sound-absorbing material.

Factors controlling sound insulation

Transmission loss or sound reduction index. The transmission loss of a panel refers to the ability of the panel to resist the transmission of energy from one side of the panel to the other. When sound energy falls on a panel most of it is reflected, but a small fraction is transmitted to form a sound wave beyond the panel. The fraction of energy transmitted is called the "transmission coefficient" τ and the transmission loss (TL) is related to τ by:

$$TL = 10 \log_{10} (1/ \tau)$$

Transmission loss is a property of a panel and is only one of a number of factors that go to make up sound insulation. This is because sound insulation is a property of adjacent rooms and includes the effects of room acoustics and flanking transmission. A number of factors control the transmission loss of a wall: these include the mass of the panel, the number of layers it contains and its structural stiffness.

Mass law. The mass law is the basic rule of thumb for determining the transmission loss of a panel.

If the mass per unit area in kilograms per square metre (M) is known, then the transmission loss at a frequency f will be given by:

$$TL = 10 \log \left[1 + \left(\frac{fM}{131}\right)^2\right]$$

The sound insulation will increase by between 5 and 6 dB per octave (one octave represents a doubling of frequency) and by a similar amount for the doubling of the mass of the wall.

Now, subjectively, a difference of 5 dB in the sound level of a noise means

Figure 3.10   Average transmission loss as a function of surface density according to the mass law

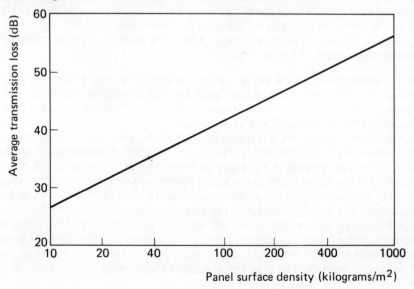

Figure 3.11   Typical sound insulation provided by a lightweight stiff panel

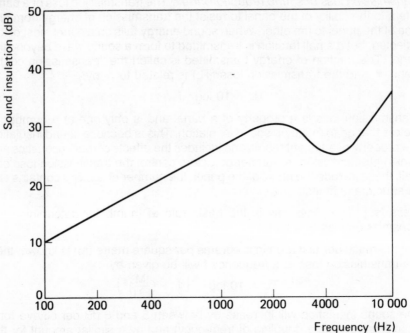

that only a small difference in the level has been obtained. Thus it is often difficult to obtain a significant increase in the sound insulation of an existing wall by merely increasing the total weight of the wall. A domestic party wall consists of a double layer of brick which can weigh up to 400 kilos per square metre (90 lbs per square foot). It is clearly not practical to add, say, 10 dB to the mass law insulation of such a wall by increasing the weight by a factor of 4. The position is more hopeful with light-weight partitions, but even so it is better to proceed on a basis of understanding of the overall problem than merely to increase the weight of the partition.

The coincidence effect. A panel can be deformed in many different ways without disturbing the edge position (boundary conditions) by placing different uneven loads on the panel. Sound waves falling on the panel have the same effect except that because the pressures due to sound patterns (modes) in the room are time-dependent, it is possible to excite a panel into a high response in two ways. A panel mode will respond well to a room mode when either the shape of the panel mode is the same as the shape of the room mode, or the frequency of the panel mode is the same as the frequency of the room mode. Normally the two effects occur separately, and it is now known that at low frequencies the modes that have resonant frequencies equal to the room mode frequency (the resonant modes) respond well and control the vibration amplitude of the panel, but these modes do not radiate sound well. The modes with a similar shape to the forcing room mode respond well (but not as well as the resonant modes), but it is these panel modes responding above their resonant frequencies, and hence in a mass-controlled manner, which couple well with the space beyond the panel and carry the sound energy through the panel. It can be shown that this results in the same mass law as was originally derived by considering infinite panels. The modes in the two-dimensional panel, however, behave differently as a function of frequency, than do the three-dimensional room modes, and as frequency increases, the resonant frequency of the shape-coupled modes becomes equal to the resonant frequency of the room modes. Thus we have a coincidence of resonant and non-resonant coupling. This coincidence leads to a sudden high coupling of the room/panel/room system and a consequent drop in the sound insulation. Typical behaviour is shown in Figure 3.11. Here we see the characteristic loss of insulation at the coincidence frequency followed by the steep recovery back towards mass law as the coincidence effect becomes less important at higher frequencies.

This coincidence behaviour is quite common in light-weight panels such as plasterboard, asbestos, glass and wood, and is often a limiting factor in the insulation that can be obtained with these materials. Thus control of the coincidence effect can result in a significant improvement in sound insulation. Much research work has been carried out by the manufacturers of these materials (particularly by the manufacturers of glass and plasterboard). These manufacturers will readily advise on the choice of suitable configurations to avoid coincidence troubles.

Multiple layers. An effective method of increasing the sound insulation of a wall is to construct the wall of a number of separate layers. Since it is the discontinuity at the boundary between two different materials that causes reflection and therefore reduction of the transmitted sound, it would appear reasonable that the more discontinuities there are the less sound will be transmitted. However, a number of theoretical and practical difficulties arise. From a theoretical point of view, sound which is reflected from one layer will travel back to the previous layer and then be reflected again. This sets up a multiple reflection between successive layers of the multi-layer wall and the overall effect is to reduce the insulation below that which might be expected from the sum of the individual insulations of the individual layers in the wall.

The introduction of gaps in a wall also permits the presence of two sorts of resonance. At low frequencies the masses of the two walls react against the springiness of the trapped air and this causes a mass spring mass resonance which in a typical multi-layer wall has a frequency between 50 and 150 Hz. At much higher frequencies, it is possible that the thickness of the air cavity could become equal to one half of the wavelength of the sound. When this happens, standing wave resonances occur in the air cavity. Both of these effects can cut down the beneficial increase of insulation derived from multi-layer panels.

Upon studying measured values of the sound insulation of such walls, it is found that the effect of the low frequency mass spring mass resonance is to hold down the sound insulation at low frequencies so that however many layers are incorporated in the wall, the low frequency sound insulation is not improved. However, the gradient of the insulation versus frequency curve is found to increase in proportion to the number of layers in the panel. Experimental measurements have been observed showing that the mass law increase of 5 dB per octave for a single panel is increased to about 10 dB per octave for a double layer panel and further to 15 dB per octave for a three-layer panel. There are some indications that this can be extended to a rather large 20 dB per octave for a four-layer panel. At high frequencies around 3000 Hz and over, the effect of the standing wave resonances can be seen. This is not a marked effect and does not even bring a reduction in sound insulation such as occurs with the coincidence effect, but generally leads to a flattening off of the sound insulation versus frequency curves which nevertheless continue upwards.

The beneficial effect of multi-layering can often be obviated by internal vibration bridges within the panel. For structural reasons it is often impossible to build a double panel without some form of cross bracing between panels either on the edge or over the face of the panel. This produces vibration bridging which can transmit sound through the panel cancelling out the beneficial effect of air gaps and other discontinuities. A case in point concerns an experiment that was carried out with steel plates containing a vacuum. The idea of this was that a vacuum should not conduct any sound and this multi-layer panel, in theory, should provide perfect insulation. In practice, to sustain a vacuum between two plates a large number of internal struts were necessary, keeping the two plates apart. The sound insulation was measured before and after the air was taken out from between the panels and it was found that the sound insulation of the panel with a vacuum was less than without. This was because the pressure of the atmosphere caused an improvement in the coupling of the two plates via bridges which then transmitted sound energy with greater efficiency.

In practice the effect of flanking has been found to overcome almost completely the beneficial effect of layering so that a double brick wall has a performance little better than it would have if it were a single layer acting according to the mass law.

Sound absorbing material. Sound absorbing material has a role to play in controlling insulation that can be obtained between two rooms. This will be discussed shortly. The present discussion applies only to sound absorbers which improve sound transmission loss of a panel. It is thought that by placing sound absorbers inside the cavity, the resonance mentioned above can be reduced and so the sound insulation of the panel increased. In practice, the results are not as good as may be expected for two reasons: in the first case, the mass spring mass resonance occurs at a very low frequency and sound absorbers do not have good sound-absorbing properties in this range. It is then found that sound absorption has little effect on mass spring mass resonance. The small amount of sound absorption that may be present already in the cavity is sufficient to drown this resonance very strongly. Hence the addition of further absorbers has little effect. Sound absorbers may also be placed over the face of the panel and if the panel is light-weight and has small insulation, quite significant improvements in sound insulation have been observed. As much as 15 dB can be obtained at high frequencies, especially where there is a coincidence effect, since the presence of the sound absorber cancels out the increase in transmission obtained by this effect. Sound absorbers by themselves have very little inherent sound insulation. A four inch layer of polyurethane foam will have transmission loss of only 5 to 8 dB.

Sound insulation and transmission loss. The quoted transmission loss of a panel is not necessarily the sound insulation that will be obtained between two rooms separated by that panel. Other factors come into play. Principally these are the exposure area (S) and the effective sound absorption within the room receiving the noise (A). Sound insulation (I) and transmission loss may be related by the equation:

$$I = TL - 10 \log S/A$$

This equation shows that the larger the exposure area between the two rooms, the smaller will be the insulation. However, if the absorption in the receiving room is increased this will increase the insulation. Insulation obtained in practice is thus generally related to the area of sound absorber, but we must note that it is inefficient to try and improve sound insulation by increasing sound absorption since normal rooms have a fair amount of sound absorption anyway, and that according to the equation, even if this amount of sound absorption were doubled, the insulation would only increase by 3 dB, which is hardly a significant increase in insulation. The quantity 'S' can only be controlled at the planning stage and shows the importance of minimum areas of contact between noisy and quiet areas.

3.7 NOISE CONTROL AT THE RECEIVER

Introduction

Noise is received by people and more exceptionally by delicate instrumentation and it is often necessary to control the level of the noise received. This is normally done by treating the room or area within which the receiver is located and we therefore have to study the acoustics of this situation.

Room acoustics

Introduction. This subject will be revised to a certain extent later. However, the main points will be mentioned. Consider a source of sound located within a room radiating sound energy. At most frequencies the sound energy travels away from the source and is multiply reflected from the boundaries of the room, and a uniformly diffuse sound field (which is called the reverberant field) builds up within the room. Figure 3.12 shows the normal situation from which we can define three areas of interest. First there is the near field within which we are so close to the source that individual characteristics of the source control the high sound level. Further away from the source there is a free field in which the source can be treated substantially as a point source at which the noise levels fall off by 6 dB for each doubling of the distance from the source. This free field can lead, but does not always, into the reverberant field which fills the rest of the room space. The reverberant field is usually uniform within the room except in the region of the source or near to highly absorbing areas.

The near and free fields. In the near and free field the sound level is controlled by the sound being directly radiated from the source and is independent of the room acoustics. Any control of the sound level therefore must be by means of direct reduction of the source power or by means of a barrier between the source and the receiver.

Free field radius. The free field radius is the distance from the sound source at

Figure 3.12 Sound fields around a source in an enclosure.

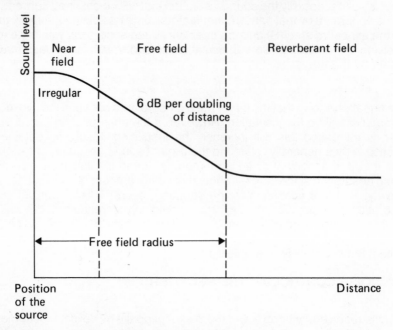

which the free field merges into the reverberant field. If the source has directional characteristics this radius will depend on orientation and it can also be affected by the presence of other sound sources within the room. It is possible for the free field radius to exceed the typical dimensions of the room, in which case an effective reverberant field does not exist within the room and free field noise control techniques only can be used.

Reverberant field. The reverberant field level is controlled by the power of the source, the size of the room and the amount of sound absorbing material within the room. Within the reverberation field it is possible to control the sound level by controlling the power from the source, for example with enclosures, or by means of sound absorbers within the room, but not by means of sound barriers (as opposed to enclosures) near to the source. By increasing the sound absorption within the room, the reverberant field level can be reduced but this is not an efficient process since it is necessary to double the area of sound absorber in order to reduce the reverberant field level by 3 dB. Most rooms already contain a fair amount of sound absorber so that the cost benefit ratio for this procedure quickly becomes uneconomic and indeed impractical. Another effect is that when the absorption in the room is increased the free field radius is also increased.

3.8 NOISE CONTROL IN LARGE ROOMS WITH MANY SOURCES

Introduction

In a large room with many sources, the model of near, free and reverberant field still applies, but there are complications. The level of the reverberant field is controlled by the total sound power that is being emitted into the room and this level can exceed the level in the near field of some of the quieter noise sources, so that zero free field radius will apply to these sources. The free field radius of a dominant source will be substantially the same as if that source alone was radiating into the room, but if a number of sources of roughly equal strength are dominant then the free field radius will be reduced by the increased reverberant field due to the other sources. Fortunately, it is quite easy to trace the extent and level of the various fields by moving a simple sound level meter around the room.

However, the problem of reducing the reverberant field level to a specific level (such as 90 dB(A) to come into line with the Code of Practice) is not straightforward in that if the dominant noise source is treated, the sound field will not necessarily fall by an amount equal to the reduction in the level of the dominant sound source. What happens is that the sound field falls to the point where the next loudest source becomes the dominant source and not much below this. A process of progressive reduction is thus initiated until enough sources have been tackled to reduce the field to a reasonable level. In all cases, the dominant source can be identified as being the one with the largest free field radius and usually the loudest near field sound level.

Pure tone effects

All the above theory applies to complex or random sounds which are produced by such sources as moving gas, motion of wheels on floors, and the like. The behaviour of pure tones is entirely different. Pure tones are produced by vortices, humming noises from electrical devices and machines, rotating fans, and resonating columns of air.

Low frequencies

At low frequencies a typical sized room will have standing wave resonances which are few in number over quite a wide frequency range. Thus the response of the room to low frequency sound will be concentrated at the discrete frequencies of these modes and at low frequencies behaviour tends to be tonal even for random noise sources. The situation is worse when the noise is itself tonal, for example the noise produced by transformers, and it is possible, depending on the precise geometry of the room, for the source to be in tune with a standing wave within the room. When this occurs, large sound pressure levels can be developed which can be unrelated to sound pressure levels in the adjacent rooms and are often excited by machines at quite remote points in the building because of vibrations passing through the structure of the building. Samples of such phenomena are fortunately rare but can be serious when they do occur.

High frequencies

At higher frequencies pure tones can be treated for noise reduction purposes much in the same way as more random sound fields, except that there are unusual subjective phenomena normally associated with these fields. Principally, when a pure tone noise fills a room it can be found that by simply moving a head a small amount, the subjective loudness of the tone will alter considerably. This is because the standing wave nature of the sound is still preserved, even though the wave length may be much shorter than the room size.

3.9 CONCLUSION

In recent years the body of knowledge about noise control has grown very rapidly. There are now in existence many professional organisations specialising in noise control and both the public and the government have become aware of the necessity to control noise and to treat it as one of the many by-products of technology that can be regarded as pollution. Although there is certainly as yet no reason for complacency, the major noise problems of aircraft, traffic, and the construction industry are being tackled. It is within the grasp of present technology to substantially reduce noise from these sources. As so often happens, it is merely a question of whether or not the money, amounting to some billions of pounds, for example, in the aircraft industry, will be spent.

4. Surface transport noise

4.1 INTRODUCTION

In this chapter we will discuss ways in which traffic noise is measured, factors that affect it, and means that can be used to control it. We shall also explain that part of the UK's Land Compensation Act (1973 Amended 1975) which enables people to obtain sound insulation against increased noise due to new road constructions.

We start by considering the legal constraints upon the sources of the noise themselves which, for example, in the UK, are to be found in the Motor Vehicle (Construction and Use) Regulations (1969 amended 1973). These regulations set emission limits for the amount of noise that individual vehicles can produce and provide the police with powers to test vehicles with noise-measuring devices in the same way as radar traps detect individuals who break the speed limits. These powers have, however, been underused.

We go on to consider how traffic noise depends upon traffic composition and speed and how the sound from the road travels towards the receiver, including the effect of ground attenuation as the sound travels through trees, across fields and other terrain.

Occasionally we have to consider the likelihood of noise nuisance from a projected road scheme, so next we discuss the methods of predicting road traffic noise. We end the chapter with a discussion of general methods for dealing with road traffic noise and some thoughts on railway noise.

4.2 CONTROL OF VEHICLE NOISE AT SOURCE

Individual vehicle noise is either associated with the vehicle's movement along a road, or with its engine and transmission system. The movement noise tends to vary with road speed, while engine and transmission noise vary with engine speed. For most urban driving conditions it is the latter noise sources that dominate. It is usually only at high road speeds, over 80 km/h, that movement noise becomes the main source.

Movement noise includes the tyre-road interaction noise, wind noise, and payload or body noise. Tyre noise increases with the cube of road speed, but it is also affected by tread design, tread wear and the road surface itself. Unfortunately, many of the tyre-road surface designs which have been developed to improve traction, and thus safety, have also been found to increase noise emission – (the quietest tyre is one with a bald surface on a smooth road).

Most European countries, Australia, Canada, and many states in the USA specify noise emission limits on road vehicles. The UK and several states in the USA specify drive-past and static (with engine revving) tests for checking noise levels.

The UK Motor Vehicle Construction and Use Regulations (1969 revised 1973) include specific emission standards in dB(A) for noise from motor vehicles as well as more general noise control requirements. Every vehicle propelled by an internal combustion engine is required to have a silencer or some other device (although the authors are not aware of an alternative) to reduce the noise from the exhaust "as far as may be reasonable". As well as this construction requirement, the way in which the vehicle is driven must be such that it does not cause "excessive noise".

In addition to these somewhat vague controls there are two checks that can be made on the noise from any vehicle driven on the roads. In the UK there is a static test for vehicles first used after 1 April 1970, in which the vehicle is mounted on rollers and taken up to two-thirds maximum engine speed in first gear. These limits are set for several categories of road vehicles, including automobiles, trucks, motorbikes and buses. This is really a production or quality control test, since not every vehicle of every make has to be tested against the specified noise limit before it is driven on the roads. Different static test limits are set for vehicles first used between 1 April 1970 and 1 November 1970, those first used between 1 November 1970 and 1 April 1973, and those first used after 1 April 1973. It is the intention to reduce these limits progressively to take account of advances in noise control technology. Currently, for new vehicles, the limits stand at 88 dB(A) for cars and 92 dB(A) for heavy vehicles (peak). (See Table 4.1).

The noise from vehicles first used before 1 April 1970 but after 1 January 1931 may be checked when moving along a road (drive-past test), but the conditions under which such measurements may be made, eg open and flat terrain, are hard to find. This is true of the American Society of Automobile Engineers test also, since both it and the British test require at least 10m of open terrain from the kerbside. Often tests can involve a specially-constructed test site, but in Chicago, Illinois, and San Francisco, California, corrections to allow for the presence of nearby reflecting surfaces in the drive-past test are provided. The moving vehicle test limits are 3 dB(A) higher than those for the static test, even though there is no specification of engine load or anything else in the moving vehicle test.

There are proposals to further reduce noise emission limits for vehicles in the UK in line with other EEC countries and to introduce a test for noise emission into the annual road test for heavy goods vehicles (trucks). The proposed limits are 80 dB(A) for automobiles and 89 dB(A) for heavy goods vehicles.

Two of the problems with the emission limit enforcement system are the load on police officers who are busy with what they consider to be more important offences and the difficulty in finding adequate locations for the site of vehicle

Table 4.1 Maximum permitted sound levels (db(A)) for road vehicles

Vehicle	Proving (BS 3539)	Road Test
Motor cycle <50cc capacity	77	80
Motor cycle >50cc <125cc	82	85
Motor cycle >125cc	86	89
Goods Vehicle, Motor Tractor Locomotive, Land Tractor Works Truck, Engineering Plant Passenger vehicle (more than 12 persons)	89	92
Passenger vehicle (less than 13 persons)	84	87
Motor Car & Any Other Vehicle	85	88

monitoring. These have been tackled by the city of Lakewood in Colorado, USA. The Environmental Control Division of the local authority maintains two monitoring vehicles which are installed in locations which satisfy simple criteria for microphone position – 1.2 m (4 ft) above the ground, at least 8 m (25 ft) from the near side of the lane being monitored, even with the roadway elevation, and at least 32 m (100 ft) from an intersection on a less than 2 per cent grade. The monitoring vehicles operate in conjunction and in communication with a chase vehicle which is installed at a distance of $\frac{1}{2}$ km (a quarter of a mile) ahead. First offenders receive a citation, but they can avoid a fine by locating and correcting the cause of excessive noisiness and checking this at a compliance centre. Similarly successful enforcement procedures are to be found in many other cities in the USA, including San Francisco.

4.3 NOISE EMISSION LIMITS IN THE USA

The Noise Control Act (1972) in the USA provides for a division of powers between federal, state and local authorities. The state and local governments retain the right to establish and enforce controls on environmental noise by regulating the use, operation or movement of noise sources and by establishing emission standards. An example of local authority action exists in Chicago, where from January 1st, 1975, motorcycles have been restricted to a maximum of 84 dB(A). With regard to automobile and truck noise, several state authorities, including the state governments in California and Florida, have taken action requiring a schedule of decreasing allowable noise levels. Indeed, California will require automobiles to meet a 75 dB(A) emission limit from 1980. Ninety-seven per cent of California's citations follow enforcement of the state laws by visual inspection and the issuing of equipment-correction citations (buy a new muffler!) rather than by noise measurement. Hawaii has included a time restriction as part of its sound level limit: trucks are limited to 78 dB(A) between 11 pm and 7 am.

The implementation of such measures is curtailed during a fuel shortage since the noise-reduction techniques necessarily introduce a reduction in engine efficiency. No state in the USA is able to introduce an emission limit on inter state trucks which is out of line with federal requirements and these are still being researched. The US Bureau of Motor Carrier Safety (BMCS) in carrying out its responsibility to enforce the federal noise emission standards for inter state motor carriers, uses the stationary test procedure, but is concerned essentially with cab noise.

A model noise control legislation including traffic noise abatement has been developed recently by the Environmental Protection Agency (EPA) and the Council of State Governments, and this will supersede individual state legislation.

4.4 METHODS OF MEASURING AND PREDICTING TRAFFIC NOISE

Methods of measuring sound have already been discussed in Chapter 2, but we will go into more details here. It is useless from the point of view of legislation to stand by a road and measure with a sound level meter the instantaneous sound level that exists or the average level over a small sample time period. The latter measurement will depend on the variation in flow of traffic and instantaneous readings will be influenced by the passage of individual vehicles. So the sound from the road has to be measured over a sufficiently long period of time to give an overall idea of the noise level. Indeed, in the USA and Japan, the average level, or L_{50}, has been used to assess highway noise annoyance. It has now been decided that L_{50} shall not be used to determine the annoyance value of the noise from the road in the UK but that a more complicated measure shall be used – that is, the eighteen-hour ten per cent level in dB(A), or the eighteen-hour L_{10} index. The eighteen-hour L_{10} index is the noise level in dB(A) that is exceeded for 10 per cent of the time between 0600 and 2400 hours GMT. The sound level during sample 15-minute intervals, one in each hour, from 6 am until midnight is measured at the roadside. A large number of instantaneous measurements are made within each 15-minute period and these are analysed in order to find out the level exceeded for 10 per cent of the time. The full eighteen hour determination is, however, only necessary when the L_{10} value is within a few dB(A) of the 68 dB(A) compensation limit. Where initial measurements show that the noise level is unlikely to be near 68 dB(A), shorter methods may be used to determine an estimate of L_{10} for the full 18-hour period (see Bibliography).

Rough values can be obtained by spot-reading a meter that reads only in dB(A). To do this, you simply note down at regular intervals (say, one every 10 seconds, although a shorter interval, 4-5 seconds is preferable), the sound level as shown on the meter. When a hundred readings have been obtained, the top ten readings, that is the highest ten levels in dB(A), are deleted. The highest of the remaining ninety levels is then roughly equivalent to the L_{10} value. A more accurate method of obtaining L_{10} is to plot the distribution of levels (level versus number of registrations at that level) and to calculate the standard deviation about the mean of the distribution. Assuming that the distribution is gaussian – normal (ie fits a bell-shaped curve which is symmetrical about the mean) then the 10 per cent level is found by adding 1.28 × the standard deviation to the mean. Similarly, the 90 per cent level (L_{90}) is found by subtracting 1.28 standard deviations from the mean (Figure 4.1). The formula for standard deviation is

$$\sigma = \sqrt{\frac{\Sigma \ (\bar{x} - x_i)^2}{N^{-1}}}$$

where $\bar{x}$ is the arithmetic mean of the total of N readings and x_i is the ith reading.

Figure 4.1   Normal distribution curve showing 1.28 standard deviations either side of the mean

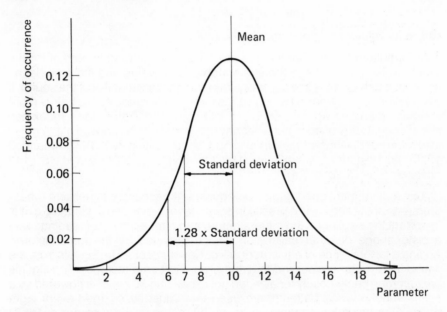

Needless to say, this is a very tedious method of calculating L_{10} and requires considerable dedication, especially when it has to be done eighteen times. An equally accurate but substantially less time-consuming method of calculating L_{10} from spot readings is again based on the assumption of a normal distribution. It is necessary to sub-divide the total range of noise levels into consecutive 2.5 or 5 dB(A) ranges and to compute the fractional cumulative probabilities in each 2.5 dB(A) or 5 dB(A) level range. These can either be plotted on cumulative-probability graph paper or the probability function values corresponding to each fractional cumulative probability can be read from statistical tables and plotted on ordinary graph paper.

As remarked earlier, more sophisticated apparatus may be used to obtain the instantaneous registrations of level than the sharp-eye-with-a-sound-level-meter. A combination of a graphic level recorder and a statistical analyser can be used, but still needs a cumulative probability plot to yield L_{10}. Cumulative graph paper is difficult to obtain, but is manufactured by Chartwell and is supplied by them under their code number 5571. Most large stationers will supply this, usually on order. Recent developments in instrumentation enable the whole procedure to be carried out automatically, yielding a digital read-out of the 18-hour L_{10} in dB(A)

Having obtained our L_{10} value, we have a measure that approximates the peak noise that can be expected from the road. A measure of the steady noise contributed by the general (remote) traffic stream is the L_{90} value – that is, the value that is exceeded for 90 per cent of the time. This value must be lower than that of L_{10} and approximates the steady background noise level present during the readings due to the local distribution of noise sources. In the UK there is a composite unit called the Traffic Noise Index (TNI) which is given by the formula:

$$TNI = L_{90} + 4 (L_{10} - L_{90}) - 30$$

where all values are measured in dB(A).

This formula indicates that two variables were considered important for the Traffic Noise Index: one is the straightforward L_{90} value and the other is the difference between L_{10} and L_{90}, which represents the variability of the sound. It was reckoned that a sound which varies over a wide range of decibels is more annoying than one which is constant with time. The Traffic Noise Index has never been widely used in noise standards, and for the purposes of the noise insulation regulations of the UK's Land Compensation Act (1973 Amended 1975) the 18-hour L_{10} value is used exclusively; L_{90} is still used in BS 4142 however (see Chapter 7.4).

As we have mentioned, the basic equipment necessary to perform measurements of L_{10} and L_{90} is a precision sound level meter and a windshield. It is important to re-calibrate the meters before and after each set of readings with a pistonphone or other calibration device provided by the manufacturer. Adding a further piece of equipment will cut labour costs. The signals from the sound level meter are passed to a sampling chart data recorder. The signals are recorded continuously for a period not exceeding 30 hours, if powered by a mains supply. When batteries are used they must be charged every eight hours. One noise level reading is taken by the sampling recorder every $3\frac{3}{4}$ seconds. The range is limited to 20 dB and the precise peaks and troughs of the noise level are not recorded. Once it is set up, particularly with the insertion of a mains control switch which will automatically start sampling at 0600 hrs and cut out at 2400 hrs, and by a sampling device (such as used in air-pollution sampling) which enables a 15-minute sample to be taken every hour throughout the measuring period, the system is completely automatic and readout is accurate to ± 2 dB(A). This is sufficiently accurate for measurement of the L_{10} index as required by Noise Insulation Regulations of the UK's Land Compensation Act. However, a more widely-used system passes the signals from the external microphone through a precision sound level meter and into a tape recorder. The tape is then processed subsequently by passing the signals back through the precision sound level meter (now used as an attenuator) into a pen recorder and then to a statistical distribution analyser. From the statistical distribution analyser it is necessary to plot the 12 channel readings onto probability paper before deducing the required value from the graph. This needs experience. Furthermore, the accuracy of the tape recorders as a means of recording audible sound for later analysis is in question. As such, the tape recorder used in this way is a system weakness. If the weather is very cold or very humid, the accuracy of the tape recorder can be affected. Nevertheless, this system enables measurement of L_{10}, L_{50}, L_{90} or L_{EQ} with reasonable accuracy given fair weather conditions.

More recently developed alternative systems can record levels from an external microphone onto a cassette tape recorder via a data logger in digital form for subsequent or real-time processing by a mini-computer. These systems are good and accurate without necessariy being more expensive. They are particularly suitable for measuring L_{EQ} and L_{DN} defined by:

$$L_{EQ} = 10 \log_{10} \left[\sum_{i=1}^{\Sigma} \frac{f_i}{100} \left(10^{L_A/10} \right) \right]$$

and $\qquad L_{DN} = 10 \log_{10} \left\{ 1/24 \left[15 \left(10^{L_D/10} \right) + 9 \left(10^{\frac{L_N + 10}{10}} \right) \right] \right\}$

where f_i represents the fraction of time (expressed as a percentage) in the i-th sound level class (usually a 5 dB(A) interval, eg 45 to 50 dB(A) or 65 to 70 dB(A)),

L_A is the sound pressure ratio corresponding to a particular dB(A) level,

L_D is the L_{EQ} for the daytime (07.00 to 22.00 hours) period,

L_N is the L_{EQ} for the night time (22.00 to 07.00 hours) period.

Note that in the definition of L_{DN} the night-time period has a 10 dB penalty.

4.5 PROPAGATION OF ROAD TRAFFIC NOISE FROM SOURCE TO RECEIVER

In many cases it is important to know the noise levels that will be generated about a road before it is built. It is also tedious to carry out a measurement at every point of interest around an existing road. So there must be a method of calculating the noise level at any point which takes into account: distance from the nearest kerb, traffic flow and composition, ground attenuation, barriers, elevations, and cuttings. In addition, with respect to the insulation of the dwellings, their orientation, sound insulation of windows and doors, and acoustics of the home environment need to be noted. This gives a good indication of the likely noise nuisance from a road. In the USA predicted noise levels to be used in assessing noise impact must be obtained from a prediction method approved by the Federal Highway Administration. The procedures that can be used by highway engineers and planners and anybody else who wishes to use them in the UK are laid out in "Calculation of Traffic Noise", a technical memorandum published by the Department of the Environment (see Bibliography).

This comprehensive document is based partly on sets of measurements carried out by the Building Research Establishment using the motorway noise from the M1 as a source and on a set of artificial barriers and microphones placed at different distances from the motorway as a means of measuring the noise, and partly upon work carried out at the National Physical Laboratory. The initial requirements of the method are knowledge of the probable traffic density, the percentage of heavy lorries and the gradient of the road. Graphs can then be used to deduce likely 18-hour L_{10} at 10 metres. For accurate work the document itself must be consulted, but the main results are summarised here.

Basically, free-flowing traffic with a density of 10 000 vehicles per 18-day will produce an 18-hour L_{10} of 68.0 dB(A) at 10 m from the kerbside of a level road. This will increase by 3 dB(A) for every doubling of the traffic density. This

Figure 4.2 Variation of L₁₀ index with vehicle flow

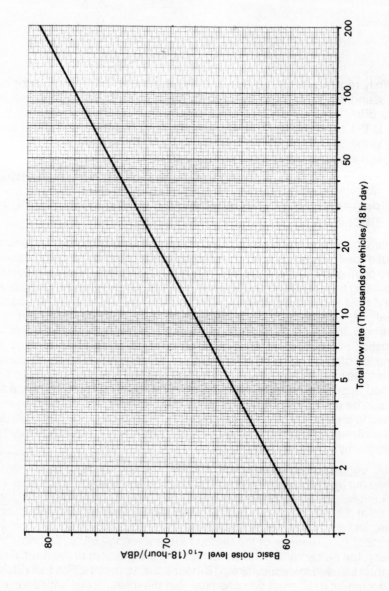

Figure 4.3   Correction for mean traffic speed and percentage of heavy vehicles

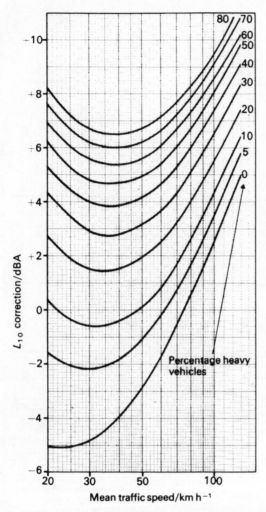

assumes that the mean speed of the traffic is 75 kilometres per hour and that there are no heavy vehicles (Figure 4.2). A correction should be added to this value, varying from minus 3 dB(A) for a mean speed of 50 kilometres per hour, rising to plus 2.5 dB(A) for a mean speed of 100 kilometres per hour and no heavy vehicles in either case (Figure 4.3).

For the purpose of these calculations, the mean speed of the traffic is taken essentially to be equal to the speed limit on the road, and detailed variations are taken according to the type of road, eg number of carriageways. On level sites corrections may be made for the distance from the road (Figure 4.4). Obviously, 10 m from the edge of the nearside carriageway the correction is zero, since this is the distance for which we have predicted our basic 18-hour L_{10} anyway. For every doubling of this distance over hard ground, we subtract

49

Figure 4.4 Correction for horizontal distance over hard ground and height above the ground

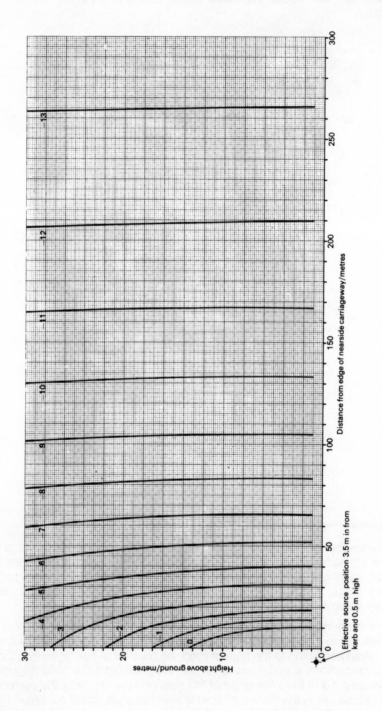

Height above ground/metres

Distance from edge of nearside carriageway/metres

Effective source position 3.5 m in from kerb and 0.5 m high

Figure 4.5 Correction for horizontal distance over grassland and for height above ground

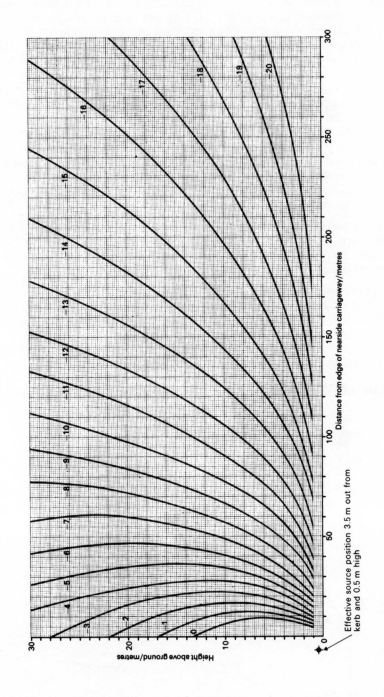

Height above ground/metres

Distance from edge of nearside carriageway/metres

Effective source position 3.5 m out from kerb and 0.5 m high

51

3 dB(A) irrespective of height. This is because the road is treated as a line source rather than as a point source. If the sound is travelling over soft ground ie grass or anything that is not paved, then the correction factor is larger and more complicated since it depends upon height and to some extent upon the nature of the ground cover. For example (Figure 4.5), if the average height of propagation above grass-covered ground is 5 m over a distance of 100 m, the correction is −13 dB(A). Allowance can also be made for the distance down side roads and for reflections due to the presence of other buildings. Finally, to calculate the level at 1 m from the facade of a noise-exposed building as required by the regulations, a correction of +2.5 dB(A) must be made due to reflection from the facade. By reflection from the facade, we mean a kind of instantaneous echo which increases the noise level in front of the facade. Strictly, the amount of this increase depends upon the nature of the facade and other factors; however, a constant correction is applied in the UK government memorandum.

All these procedures enable the calculation of the probable 18-hour L_{10} noise level at a nearby facade. However, if the 18-hour L_{10} exceeds the compensation limit (68 dB(A)), then we need to know how to reduce the noise level. This can be done with noise barriers.

Performance of barriers
Specially constructed noise barriers are a frequently-adopted method of controlling noise from major highways in America and Europe. However, acoustic barriers are not completely effective in eliminating sound. In fact, a barrier of a reasonable height (not more than 5 m) is limited to a maximum attenuation of 20 dB(A), however massive and thick it is. To understand this, a distinction

Figure 4.6   The effect of a noise barrier on facade L_{10} and L_{10} minus L_{90}

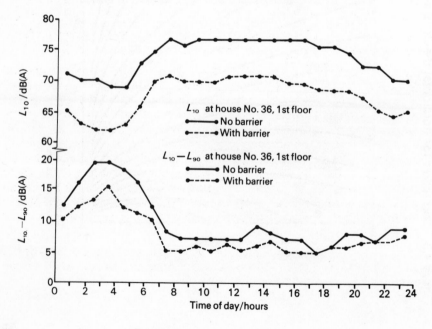

should be made between a barrier and a partition. A barrier is a construction which cuts the line of sound between source and receiver, but which allows sound to bend around it like water waves around a breakwater. A partition,for example an airtight partition between two rooms, allows no form of airborne path whatsoever. It is the bending of sound around the top and the ends of a barrier that limits its attenuation. So the effectiveness of a noise barrier is dependent upon, among other things, the length of the diffracted noise path compared with the line-of-sight distance between the source and the receiver.

Figure 4.6 shows the results of the Building Research Establishment's research on the performance of a noise barrier erected near the M4 in the UK. In this case, the barrier is made of hollow vinyl panels 2.7 m high and 300 m long, with a weight per unit area of 7.7 kg/m^2, but effective noise barriers can consist of buildings, banks or mounds of earth, hillocks – in fact any impervious structure (ie without air gaps) that is erected between the road and the observation point.

The use of noise-insensitive buildings such as warehouses or factories to screen sensitive buildings such as houses, schools, offices, hospitals etc beyond them is sensible planning policy.

A common error is to assume that a clump of trees or a thin line of under-growth is an effective barrier. The amount of sound attenuation obtained by a 50 m thick, dense, evergreen forest with foliage down to the ground level can be between 3 and 5 dB(A). The fact that most people think that screening a highway or other noise source with trees reduces the noise level is due to the psychological effect of making a source invisible and therefore not readily identifiable. This tends to reduce the overall impact of the noise source on the individuals concerned, even though they are receiving only a little less noise. From this point of view, the combination of an artificial barrier with a vegetative screen is most effective. Other problems with barriers in severe climates are drifting snow and shadowing.

The prediction techniques outlined earlier in this chapter include the effects of purpose-built noise barriers and elevated or depressed highways.

In the USA techniques and practices in highway noise barrier design and construction are still evolving. However, several states have considerable experience. Materials that have been used include timber, solid concrete and hollow concrete blocks. Since a location close to the receiver is probably less desirable to the owner of the abutting property and since additional right-of-way acquisition may be required, the tendency has been to locate highway noise barriers nearer the through traffic lanes. Distances from source to barrier vary from 3 to 24 m (10 to 80 ft) with the average being about 10 m (34 ft). Earth banks which have flat or concave slopes have been constructed to offer an increased safety factor to out-of-control vehicles. Indeed, where right-of-way width and excavated material have been available earth banks typically result in the least initial cost.

State highway authorities in the USA may receive federal aid towards the erection of noise barriers where a suitable case has been made. Such aid can extend to 90 per cent of the cost of the barrier.

4.6 THE LAND COMPENSATION ACT: NOISE INSULATION REGULA-TIONS (1973 amended 1975)

In most countries, the power of local authorities to control highway noise is limited to the design and planning stage. However, recent legislation in the UK has set a remarkable precedent. The Land Compensation Act 1973 makes it compulsory for local authorities to insulate people affected by noise above a certain level (and after a certain increase in level) from new or altered roads – a palliative for inconvenience caused by the noise.

Basically, if as a consequence of the construction of a new road the level of the noise at 1 m from a dwelling house is likely to increase by one or more dB(A) to a value in excess of 68 dB(A) for the 18 hour L_{10}, then the occupant is entitled to insulation. The Act does not apply to increase in traffic along any existing roads, nor to such trivial matters as resurfacing of existing roads. However, new slipways to new roads are included. The Act also covers noise caused during the construction of a new carriageway. If, in the opinion of the appropriate highway authority, there will be serious effects over a substantial period of time, there will be a duty to give compensation or to make a grant in respect of the cost of carrying out insulation work to the building. This provision is extremely vague and requires the highway authority to be judge, jury and tribunal. It seems unlikely, given the time scale of making appeals to local authorities, that anyone would receive any compensation or insulation against road construction noise as a result of this provision. Because of the fact that there is cash involved, everybody in the UK who is subjected to new road works would do well to study very carefully the sound insulation regulations that are in force as a result of the Land Compensation Act. A local authority has a duty to carry out insulation work *or* to make grants to houses affected by noise. However, there are exclusions. If the noise level before the operation of the new motorway or carriageway already exceeded 68 dB(A) the house will not be entitled to compensation. Also, older highways which essentially become 'feeder' roads to a new highway are excluded.

The action that a local authority has to take is as follows:

1) It is required to calculate the relevant noise levels in relation to each eligible building and prepare and publish a map or list identifying every eligible building within an area likely to be subject to noise levels that satisfy the requirements of the regulations.

2) The map or list must be prepared and published not later than 6 months after the relevant date – usually the date on which the highway is brought into operation – and must be deposited at the offices of an appropriate highway authority or its local agent where it should be available for inspection during office hours.

3) In calculating the noise level at the facade of any eligible building for the purposes of these regulations a highway authority must use the specified prediction method unless, in its opinion, it is inappropriate in the circumstances of the case. Usually the prediction method will have to be used since the road will not often be open.

These regulations apply to only certain forms of building. These are:

"dwellings, and other buildings used for residential purposes but not any

building in respect of which a compulsory purchase order is in force or any building likely to be acquired compulsorily in local or private acts of parliament, or any building which is subject to a demolition or closing order, or any building within an area declared to be a clearance area, or any building which was first occupied by the current occupant after the relevant date. Also any building or part of a building in respect of which part a grant has been paid or is payable in respect of carrying out insulation work under any enactment other than the Land Compensation Act or any regulations made under any other such act".

The most important point here is that any building which was first occupied after the relevant date is not entitled to compensation. If you move into a house by a new motorway, after the opening of the motorway, it is assumed that you are aware of the nuisance caused by the noise from the roadway and that presumably you were able to negotiate a lower price for the dwelling from the vendor as a result of this nuisance. If you did not, then you were unwise, and the principle of *caveat emptor* applies.

To quote further from the regulations:–

"As soon as the highway authority have published the map or list mentioned above, they should make an offer in writing to the occupier or the person entitled to occupy every eligible building identified. This offer in writing shall identify the building to which it relates and it will be an offer to carry out insulation work or make a grant in respect of the cost of carrying out such insulation work in or to every eligible room in the building, describe the work to be carried out for this purpose and request the person to whom the offer is made, if he is not the owner of the building to notify his immediate landlord or licensee of the terms of the offer. The local authority will also set out the conditions subject to which the offer is made. The person receiving such an offer and entitled to receive it, shall in writing, accept the offer to carry out sound insulation work to some or all of the room in respect of which the offer was made, or accept the cash grant that was offered".

It should be noted that the acceptance of a cash grant does not necessarily bind the householder to use the money for the purposes of sound insulation. If instead of doing this, he buys himself a new motor car, he should remember that when he comes to sell his house the person buying it will not receive the benefit of the sound insulation work and nor will he be able to claim again against the local authority because of the provision that the person affected shall be in possession or occupation of the house before the relevant date. However, the giving of 'no strings' cash grants does not appear popular and the indications are that only marginal cases are being dealt with in this way.

The owner of the noise-affected building must accept the insulation offer within six months. However, the appropriate highway authority or its agent may extend the time limit to 12 months. The money involved is governed by the following statement:

"The amount of the grant shall be equal to the actual costs incurred by the claimant or the reasonable cost which would have been incurred by the highway authority in carrying out in accordance with the relevant specifications the insulation work in respect of which the claimant has accepted an offered cash grant whichever shall be the less".

These broadly are the outlines of the Land Compensation Act and anyone in the UK who has a new highway opened nearby should get in touch with the Local Authority immediately to find out what is being done under the Land Compensation Act.

It should be noted also that the Land Compensation Act does not just apply to noise from highways. It applies to other sources of noise, including aircraft noise due to increased noise operation of new or extended runways and to other forms of pollution or disturbance that affect property. However, no clear regulations are yet in existence concerning these matters. The fact is that only new or altered roads are regarded as sources for the purposes of awards. There is, however, some pressure to extend these definitions and early in 1978 a Member of Parliament made some attempt to introduce a bill to compensate those living near to existing motorways.

4.7 DOUBLE GLAZING

There is no doubt that double glazing increases the amount of sound insulation between dwellings and the outside world. In fact, if a building is eligible, the noise insulation regulations in the UK make provision for:

1) the replacement by or conversion to double windows of all windows in the facade and venetian blinds between the panes to control overheating due to 'greenhouse effects'
2) the provision and installation of a ventilation unit (powered inlet fan) in each facade-connected room, together with an air supply duct for supplying fresh air to the ventilator unit from outside (blocking up the existing air brick in the external wall) and having a sound-attenuating lining.

The mechanical ventilation plant is needed so that the windows can be kept closed. There is no use in double glazing windows against sound if they can be left open. An open window transmits sound almost as readily as a hole in the wall and the same is true of open doors and air bricks. The effectiveness of a wall in reducing noise is determined by the portion of its area which is window. When sound is incident upon a wall some sound is absorbed within the wall, some is reflected and some is transmitted through the wall. So the portion that is reflected depends upon the portions that are absorbed and transmitted. A point to note however, is that good absorbers are usually poor reflectors but good transmitters, and vice versa.

The proportion of incident sound that is transmitted also depends upon the frequency. It is usually high at low frequencies and small at high frequencies. In general, the proportion of sound transmitted also depends upon the mass of the wall, which is why windows are a limiting factor. More specifically the sound transmission can be related to weight per unit area (called 'superficial weight' by builders). The difference between the sound energy (or sound pressure) on one side of the wall and that on the other at one particular frequency is expressed as a ratio in dB and is called the sound reduction factor or the sound transmission loss. If frequency is doubled, the sound reduction factor will increase by approximately 5 dB. Rather than specifiying the sound reduction factor at all frequencies for any wall, it is customary to give a single figure called the sound reduction which is roughly the arithmetic average of the sound reduction factor at 16 frequencies between 100 Hz and 3150 Hz.

As well as giving extra sound reduction due to the mass of the extra pane or panel, double glazing or a cavity wall construction tends to trap sound between the two elements at certain frequencies. So if the cavity is large enough and lined with absorbing material, the resulting sound reduction can be much larger than that for a single element. You can appreciate the need for an airtight construction (see also Chapter 3).

4.8   RAILWAY NOISE

Although social surveys – for example, the Wilson Committee Report (see Bibliography) – demonstrate that noise from railways is significantly less bothersome than that from road traffic or domestic sources, the increasing speeds and frequencies employed by rail services and the increasing practice of using land near existing railways for residential development warrant some attention from the point of noise control. Public co-operation was in the past probably due to:

1)   low ranking of noise compared with dirt and smell
2)   drastic reductions in rail services in recent years

Apart from motive power noise, which is significant only at low speeds, the noise from a moving train results primarily from rail-wheel contact in which the force required for traction and guidance is transformed into vibrations which generate noise. Rail joints and wear of wheels also contribute. Very little acoustic energy is transmitted to coach or wagon structure – the primary radiation is from the wheels, and hence reductions are possible through shielding wheels. Axle load affects the noise considerably. Braking is another important source. With diesel-engined trains, noise from the engine is predominant, except when the listener is alongside the fast-moving coaches or trucks.

Noise/speed correlation is clear with passenger trains but not so well-established for goods trains. Levels can be adequately characterised by dB(A). A recent study carried out in the UK suggests that L_{EQ} (24 hours) is better (although not well) correlated with annoyance from railway noise than units or indices which take into account both noise level and number of events (see for example NNI – Chapter 5.2).

There seems to be tentative agreement that housing should not be constructed where the L_{EQ} (24 hour) due to a nearby railway line exceeds, or is likely to exceed, 60 dB(A). At this level the aforementioned survey indicates that less than 20 per cent of a population are severely disturbed by railway noise.

Table 4.2 gives L_{EQ} values at various distances from a railway line carrying approximately 120 trains per day for a typical mix of high speed diesel-hauled passenger trains and various freight trains.

These figures assume an open site where the only attenuating factor, apart from geometrical spreading, is ground absorption (refer to Section 4.5). Under these circumstances the 60 dB(A) guideline would make any housing

57

Table 4.2   Noise levels from a railway line carrying a typical (UK) mix of traffic.

Distance from track/m	25	50	100	200
24 hr L_{EQ} dB(A)	67	64	59	54

development within 100 m of the track inadvisable. In general, these levels would be enhanced near to an embankment and reduced somewhat by a cutting and wherever the line of sight between observer and wheels was broken.

4.9   CONCLUSION

What to do about road traffic noise
Although the individual can control noise from his own vehicle by adequate maintenance, by fitting a suitable silencer and even by using noise as a criterion when purchasing the vehicle, it is obviously not possible for him to deal with the noise from every other vehicle. Every vehicle on the road is required by law to produce less than a certain level of noise, so we also need adequate enforcement of the vehicle construction and use regulations. Noise from the majority of roads can also be controlled by local pressure groups, but the process is long and requires perseverence and is not easy when the road already exists. If someone really does find traffic noise intolerable often the only solution is to move away from the noise. Except in extreme cases, such people should be able to sell their houses because a surprising number of people like the noise and bustle of traffic. If one does not wish to move however, and is only mildly affected by the noise, double glazing on windows facing the road will reduce the level by up to 20 dB(A). This is an appreciable degree of quietening but is by no means silence, so it is not wise to accept any extravagant claims made by double glazing salesmen; alternatively see if they will put their claims in writing and then if necessary prosecute them under the relevant fair trading legislation – for example the Retail Trades Description Act 1965 in the UK.

If you are within sight of any new road works or road widening scheme in the UK, you should be alert to the Land Compensation Act. Find out what is being done about this and press for an insulation grant if you are eligible.

Keep an eye open for traffic management schemes since they will often cause traffic to be directed along roads that were never intended to carry such a flow – with consequent noise problems. It is also worth noting that from the noise point of view, it is better to concentrate traffic along main roads (that are already noisy) than to distribute between parallel roads.

A great deal can be achieved by distinguishing between noise-sensitive buildings or rooms. Schools, hospitals and concert halls all come into the category of noise-sensitive. Often they can be screened by factories, gymnasia etc. which are noise-insensitive.

Within a building the same idea can be used whereby noise-insensitive rooms like kitchens, bathrooms, halls etc. can be used to screen noise sensitive rooms like lounges, lecture theatres, and so on.

Finally, remember that the use of a road vehicle is a responsibility and do not allow your own vehicle to stand warming up in the road, thus causing a noise nuisance. Drive in a manner that includes noise reduction as a part of the consideration that you show for others when using the road.

5. Aircraft noise

5.1 INTRODUCTION

Aircraft are probably the most dramatic of the man-made noise sources that are heard by the general community, particularly in the vicinity of an airport. Although the number and power of aircraft increased gradually in the 1940s and 1950s, it was not until the introduction of jet-powered engines that their potential noise nuisance became widely recognised. In both the UK and the USA major objective surveys of aircraft noise levels around airports and companion social surveys of the effects of noise on nearby residents have been undertaken.

There are approximately 200 000 people living around Heathrow (London) Airport, who are moderately or seriously annoyed by airport noise. No estimate is available of the numbers of people affected by aircraft noise around major and minor airports and airforce bases in the United Kingdom. In most surveys of noise annoyance, aircraft noise runs a close second to traffic noise in terms of the number of people affected and the extent to which they are annoyed. In the USA this situation is often reversed.

Since the purpose of an aircraft engine is to take in air and then generate a thrust by throwing out air behind it, and since noise is generated by the interaction between an ejected airstream and the surrounding atmosphere, it is not surprising that an aircraft engine is noisy. However, since the mechanical energy which appears as sound is only a very small fraction of the total energy produced by the engine, it is very difficult to quieten. Despite this, most of the major airport authorities and the airlines take the problem of aircraft noise very seriously. Their interest has been accelerated since noise certification in the UK and USA, whereby every *new* aircraft of an existing type manufactured after 1976 in the UK and after 1973 in the USA and every aircraft *of a completely new type* manufactured after 1969, must satisfy the requirements of a quality control test with regard to noise emission.

To begin to discuss how to keep out aircraft noise from a home, or how to influence the take-off procedure and routing of aircraft so as to alter the noise nuisance or similar problems, it is necessary to look at the units which measure aircraft noise. Subsequently, methods of control of aircraft noise including insulation, aircraft and airport operating procedures, planning controls and aircraft design are examined.

5.2 MEASUREMENT OF AIRCRAFT NOISE

Some experiments carried out in the USA in the late 1930s produced a number of equal loudness contours. People were subjected to a reference sound, of a pure tone of 1000 Hz, and asked to indicate when a number of other sounds of various powers and frequencies sounded as loud as the reference. However, in 1943 some other experiments in the Harvard Psycho-acoustic Laboratory revealed that high frequencies tended to be more annoying than low frequencies, even when they were judged to be equally loud. In other words, curves of equal annoyance were found which differed from the equal loudness contours. This was considered to be particularly relevant to aircraft noise, since it contains a number of very annoying high frequencies. As a result, a new unit of noise called the Perceived Noise Decibel (PNdB) was invented in the late 1950s. Various corrections can be applied to this basic unit to take account of particular pure tones in an aircraft noise spectrum, and also to take account of the duration of a typical aircraft over-flight. The level of noise created by an aircraft flying nearby increases to a peak before dying away again. Usually the length of time it takes to increase to the peak is less than the length of time it takes to die away. The duration and tone corrected form of the PNdB, called the Effective Perceived Noise Decibel (EPNdB), is used in the noise certification procedure.

The extent to which anybody is annoyed by aircraft noise will depend upon the activities which are interrupted, upon the time at which the interruption occurs, upon the frequency with which the interruptions occur, and upon his attitude towards the source. You will probably be as annoyed by one aircraft which awakens you during the night, as you will be by several aircraft flying over during the day when you are listening to the radio or doing some other activity. The Noise and Number Index (NNI) is a unit which takes into account some of these variables and which is based upon a survey of annoyance with aircraft noise carried out around Heathrow Airport in 1961. The most important factors in this index are the average peak level of aircraft noise in PNdB and the number of planes heard to be creating a level of noise above 80 PNdB (N) at some point of interest around the airport. The formula for calculating NNI is:

$$NNI = \text{Average peak level in PNdB} + 15 \log_{10}N - 80.$$

An alternative to NNI which uses EPNdB in place of PNdB is known as the Noise Exposure Forecast (NEF). The formula is slightly different in this case:

$$NEF = \text{Average peak EPNdB} + 10 \log_{10}N - K$$

where K = 88 for day-time exposure and K = 76 for night-time.

The relationship between NNI and annoyance is shown in Figure 5.1. The annoyance scores were calculated according to the answers to certain key questions; for example, one point was added to the score if the respondent was annoyed while watching television, another point was added if aircraft noise wakened the respondent or made his house shake or vibrate etc. The categories of annoyance were self-rated. NNI is calculated for aircraft movements during the 12 day-time hours between 6 am and 6 pm, or for the 8 night-time hours between 10 pm and 6 am. The night-time figure is multiplied by 3/2 to account for the missing 4 hours (6 pm to 10 pm).

Figure 5.1   Plot of NNI against annoyance (including both annoyance scores and self-rated categories)

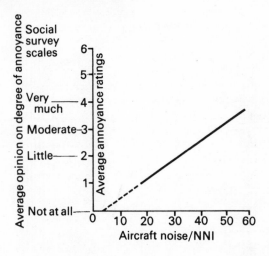

The second survey of aircraft noise around Heathrow Airport in 1971, relating to an area 30 miles by 20 miles, did not invalidate the conclusions of the 1960 survey. Similar later smaller surveys at other UK Airports tend to confirm those findings also.

From knowledge of all the important parameters like types of aircraft, their noise 'footprints', frequency of movement and routing (including runway usage) and local ground characteristics, it is possible to predict NNI or NEF contours around any airport.

Noise from individual aircraft, in PNdB, can be measured directly by a sound level meter provided with 'D' weighting and adding 6 or 7 to the displayed peak level. Alternatively, if the only available weighting is 'A' weighting, to the level in dB(A) should be added 12 – 14 dB to obtain an approximate level in PNdB depending on the type of aircraft.

5.3   INSULATION FROM AIRCRAFT NOISE

To insulate against aircraft noise it is possible to use the traditional methods that are effective against traffic noise, such as acoustic double glazing and replacing air bricks with specially constructed mechanical ventilation systems. But this is not enough since aircraft noise will come through the roof. Figure 5.2 shows typical noise reductions along various paths that sound can take into a building. Note that although these specify noise reduction in dB(A), similar figures would apply for reduction in PNdB. The roof space can be insulated by laying lead sheeting or a dense mineral wool or sand over the ceiling joists of upper rooms, but care must be taken not to overload the structure. It is also important to reduce noise transmission through chimney flues. Measurements have shown that with good double windows and a sound-attenuating ventilator unit, the first-floor rooms of houses with adequate walls can have good insulation against external noise of 35 to 40 dB and possibly more, without loss

Figure 5.2   Typical sound insulation provided by various parts of a house

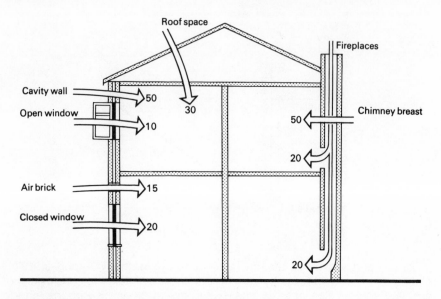

of ventilation. Rooms where open fires are in use appear to present an insuperable problem.

Needless to say, all of this is very expensive. In the UK it is possible to obtain part of the cost of insulation against aircraft noise. Grant schemes for up to 100 per cent of the cost have operated around Heathrow and Gatwick Airports, and new schemes came into operation on 1st April 1980. The British Airports Authority has also introduced schemes at four Scottish airports. Local authorities at Manchester, Birmingham and Luton operate schemes. To be eligible for one of these schemes, a house must be within the area between the 50 and the 60 NNI contour, and must have been there when the relevant scheme started. Regions greater than 60 NNI are not considered, since it is unlikely that such regions are occupied. Grants will be automatic if these conditions are met, upon application to the local authorities. Around some UK airports it is possible to obtain a rate rebate on grounds of noise exposure.

Finally, if it is possible to prove that a house has depreciated by more than £50 as a result of exposure to aircraft noise, it is sometimes possible to claim for compensation under the Land Compensation Act 1973 amended 1975.

5.4   ROUTING, TAKE-OFF AND LANDING RESTRICTIONS

Routing has a considerable influence on the pattern of noise exposure. The question is whether it is better to spread the noise burden uniformly around the airport by 'umbrella' take-off patterns, or to concentrate traffic along routes that avoid the main built-up areas.

So far, around Gatwick and Heathrow airports, the official policy has been to concentrate routing. The resulting air corridors are called Minimum Noise

Figure 5.3 Standard take-off procedure at Heathrow (London, U.K.) airport

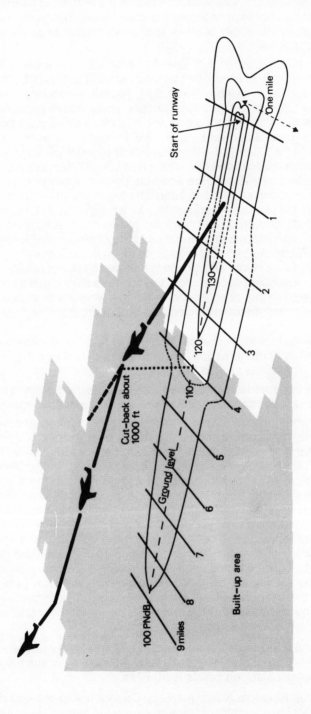

Start of runway

One mile

130

120

110

Cut-back about
1000 ft

Ground level

100 PNdB

9 miles

Built-up area

1
2
3
4
5
6
7
8

Routes (MNRs) but they attempt to minimise the number of people affected by aircraft noise rather than reducing the sound itself. This policy relies, among other things, on the ability of pilots to observe marker beacons and maintain the defined courses.

Another noise reduction measure that depends upon pilots is the specified take-off procedure. At Heathrow this requires that the aircraft be throttled back after gaining 300 m height (Figure 5.3). Aircraft are monitored at two fixed points after take-off. Aircraft that produce noise above an agreed limit are reported to the airlines concerned but little else occurs as a consequence. Heathrow's noise limits were set in 1958 in relation to large piston-engined aircraft (110 PNdB during the day, 102 PNdB by night). The same limits apply to the bigger and more powerful aircraft now flying. The implications of take-off restrictions for some aircraft are limits on fuel and numbers of passengers. A modified monitoring system has been introduced at Gatwick and others are in preparation at Luton and Manchester. As a result of a joint government and industry study and proposals from the International Air Traffic Association (IATA), the airlines have modified their take-off procedures for application at all major airports. Aircraft now employ nearly maximum power for the first 500 m (1500 ft) of their climb. This phase is followed by power reduction to normal climb and subsequent acceleration and flap retraction. Such a procedure can be compared with that described for Heathrow whereby an earlier and greater reduction in power meant a consequent stronger re-application of power later on.

This new procedure implies a decrease in noise over considerable areas some distance from the airport, offset to some extent by a slight increase in noise in a smaller area close to the airport.

Currently accepted landing routines use a $2\frac{1}{2}$ degree descent angle. Three degrees is safe for approaches at most United Kingdom airports.

A continuous gradual descent in which undercarriage and flaps are put down at the last possible moment enables high airspeeds and avoids the necessity for noisy engine power to overcome drag and to alter the slope of descent. This procedure gives distinct benefits under the glide path at large distances from the airport between 10 and 15 miles in the case of London (Heathrow). However a continuous descent procedure has disadvantages for people living under the glide path near the airport (within 4 miles in the case of Heathrow) while the aircraft are descending the last 300 m. For these people a two-segment approach is better, in which the aircraft use a steeper-than-normal glide slope and make a transition to the standard (continuous) approach path in time to stablise prior to landing. However, a slightly different angle of approach has to be chosen for each type of aircraft. Furthermore the angle of slope of the upper segment has to be chosen very carefully so that the benefit gained as a result of the steeper angle is not more than offset by the greater thrust required on the lower segment.

Considerable noise can arise when stacking procedures are in operation. For example, the noise level under a large jet at the minimum height over one of the Heathrow stacking points is 90 PNdB.

The United Kingdom Noise Advisory Council has concluded that there is little scope for improvement of noise levels around airports by routing. Take-off restrictions have been criticised on grounds of safety. In fact, they have the effect of increasing noise levels at some distance from the airport because the aircraft has not been able to climb as rapidly as possible with no restriction. So, as the area affected at the greater distances is much larger than the noise-affected area in the immediate vicinity of the airport, in terms of number of people annoyed, the total nuisance can be greater with restricted take-offs.

Night flying restrictions are very effective because of increased noise-sensitivity at night. There is a total ban on night jet take-offs during the summer months at Heathrow Airport. Further restrictions on summer night jet movements have been imposed at Gatwick, Luton and Manchester.

In the United Kingdom the responsibility for government decision-making on aircraft noise lies with the Department of Trade and Industry. In the USA, the Federal Aviation Authority (FAA) is pursuing the control of aircraft noise by requiring pilots to follow similar operating procedures including take-off and approach procedures, runway procedures, minimum altitude procedures, routing over low noise-impact areas and terminal area handling. In February 1972 the FAA "Keep-'Em-High" programme was implemented. Through this programme, arriving aircraft are kept as high as possible prior to manouvering for a safe landing approach.

With the exception of Washington D.C. National Airport and Dulles International Airport, Virginia, both owned and operated by the FAA, the public airports of the United States are owned and operated by a variety of non-federal local public authorities. These authorities have the power to impose curfews or the limitation of airport operating hours and to ban the use of certain aircraft. Such was the power invoked in the recent Concorde controversy with the consequent legal wrangling concerning the overall relationship of the federal government to state and local government entities.

5.5 PLANNING CONTROLS

At all British civil airports there are consultative committees on which local government and other interested bodies are represented and which are able to advise on noise problems. Indeed under Section 8 of the Civil Aviation Act 1968 the Department of Trade and Industry is given the power to ensure adequate consultation with the airport-noise affected public.

The Department of the Environment's Circular 'Planning and Noise' 1973 recommends a set of guidelines stated in terms of NNI. Only noise-insensitive constructions such as factories and warehouses with suitable sound insulation are considered acceptable in areas where aircraft result in noise exposures of 60 NNI or above. Dwellings and hospitals with appropriate sound insulation are considered acceptable in areas where the noise exposure lies between 40 and 60 NNI. The Circular suggests that the construction of schools is undesirable where the noise exposure is 40 NNI or above, and that sound insulation of roofs and windows plus mechanical ventilation should be provided for schools when the exposure is at least 35 NNI. These guidelines have been applied around Gatwick Airport by Surrey County Council. In the USA the use of land for residential development is only recommended if the NNI is below 38 and multiple housing is compatible if the NNI is below 55.

5.6 CONTROL AT SOURCE

The Federal Aviation Regulation (FAR), Part 36, "Noise Standards: Aircraft Type Certification" issued by the Federal Aviation Authority in 1969 requires that new subsonic jet aircraft be certified by type as to allowable noise under prescribed conditions of take-off and landing. The maximum noise level at the prescribed measuring points directly under take-off and landing paths and a position perpendicular to the landing strip (sideline) is 108 EPNdB for new subsonic jet aircraft. An amendment of the Regulation in 1973 requires compliance by newly-manufactured aircraft of older types first certificated prior to the effective date of FAR 36. A further amendment in December 1974 established noise limits for small propeller-driven aircraft. New regulatory proposals for all commercial aircraft, including aircraft capable of vertical take-off and landing (VTOL) and aircraft capable of reduced or short take-off and landing (V/STOL) are in the pipeline.

Similar United Kingdom legislation is contained in The Air Navigation (Noise Certificate) Order, 1970. However, the British limits depend upon engine size and aircraft weight. Typical aircraft noise data and the comparison with maximum noise certification (NC) levels are shown in Table 5.1. This table shows that the newer wide-body subsonic jet aircraft (BAC 1-11, Tristar, DC-10-30) have little difficulty in meeting noise certification standards. This is because they are powered by high-bypass engines such as the Rolls-Royce RB 211. The idea behind the design of an aircraft engine is to generate a thrust by throwing-out air behind it. In the turbojet, the principle cause of noise is the shearing effect in the interaction between the fast moving ejected air and the relatively still air surrounding it. The noise suppressor, which was an essential part of the hush-kit retro-fitted on jet aircraft of older types in order to meet NC standards, modifies the mixing process by putting a corrugated nozzle on the jet at the exhaust end. Unfortunately these devices have a significant effect on the efficiency of a straight jet engine and they are not used on such aircraft. Instead main reliance is placed on the other major component of the hush-kit — the acoustic lining. The basic form of acoustic lining is a sandwich of a honeycomb nature between the exposed perforated surface and the backing skin. Latest test results on a Boeing 707 fitted with such a kit show a considerable reduction of 14 PNdB at the noise certification monitoring points on landing, with a small penalty in lost operating range, because of the extra mass, of only 180 nautical miles. The reduction in take-off noise is more modest. Another technique of retrofitting older types of engines is to fit a quieter front fan.

However the most significant advance in engine design from the noise point of view has been the by-pass principle. The most recent form of by-pass engines are known as turbo fans. This by-pass engine development was based initially on engine efficiency rather than on noise reduction, but one of its spin-offs has been a reduction in noise levels. The by-pass engine diverts some of the thrust to provide a cushion of slower-moving gases between the main jet, which may still be suppressed, and the surrounding air. High by-pass engines, as the name suggests, divert a larger fraction of the jet stream than the older low-by-pass engines. So, for example, one of the new by-pass engines with approximately twice the thrust of a conventional low by-pass jet engine produces substantially less noise.

Table 5.1 Airline noise data (noise level in EPNdB)

Aircraft	Gross weight/lb	Take-off noise		Sideline noise		Approach noise	
		Measured	NC*	Measured	NC*	Measured	NC*
Aerospatiale Caravelle 12	127 870	94	96	102	103.5	104	107
Aerospatiale/BAC Concorde	389 000	114	105	111	107	115	107
BAC One-eleven 475	92 000	96	95	108	103	103.5	103
BAC Super VC 10	335 000	110	104	113.5	106.5	115.0	106.5
Boeing 707-320 B/C	333 600	112	104.2	102.8	106.5	115.7	106.5
Lockheed Tristar	430 000	96	105.5	95	107	103	107
McDonnell Douglas DC-8	325 000	117	103.5	103	106.5	117	106.5
DC-10-30	555 000	104	107.5	97	108	107	108

*NC = noise level required for noise certification.

Although the noise of individual aircraft measured as side-line, take-off or landing noise can be reduced by the application of hush-kits, and design improvements such as vertical take-off can result in less noise from a given weight of aircraft, the increasing size of aircraft, eg jumbo-jets could well cancel out part of the reduction, so the net change in the noise levels from individual aircraft over the next few years might only be a few decibels. Furthermore, if the number of airline passengers continues to increase there will be little or no shift in the NNI or NEF contours around major airports. Finally, it should be noted that super-sonic transport uses the original noisy type of jet engine in order to produce the greater thrust required. The main hope with such engine designs lies probably in development of co-annular jet with a faster moving air stream surrounding a slower moving core of exhaust.

5.7 WHAT TO DO ABOUT AIRCRAFT NOISE

Once again the primary advice to those who are sensitive to aircraft noise is stay away or move away from noisy areas. Let those who have a natural tolerance or sense of identification with aircraft live in the affected area. Having said that however, there will be those who for many other reasons have to live in affected areas, or people who consider that aircraft noise is an imposition that should be reduced so as to allow them to live again in peace. Again, some may wish to move away but will not feel it fair to stand personally a loss in cash due to reduced house value because of the noise, or who cannot afford to move due to high mortgage commitments which the reduced house value may not cover. All these people have to seek means to combat the noise.

The first step is to check on the existence of local compensation schemes and schemes for the installation of double glazing etc by local authorities. Failing these it is possible to install these at your own expense and aim to recoup the cost when selling the house. Try your building society or mortgage organization for a loan to cover the costs. Remember that powered ventilation and roof insulation are usually necessary to make a double glazing scheme really effective. Also note that it is not necessary always to insulate the whole house. Have regard to exposure (windows facing the noise) and room use and occupancy. Main living rooms and bedrooms should be treated but kitchens, bathrooms, workrooms and children's rooms can be left as they are.

Next join or form the local anti-aircraft noise group and keep informed of proposed developments at the airport. An active well-informed pressure group can directly, and more often indirectly, influence events at the airport. Keep up an attack on flight procedure, monitoring and penalties for violation, night-time restrictions and ground-running and testing procedures. Schemes for new runways and runway extensions can be opposed and the use of alternative airports investigated. Make sure that you know who is ultimately responsible for these things and keep up correspondence with him at all times.

5.8 CONCLUSION

The general world recession could mean a decline in aircraft noise. The threat of a fuel crisis in the next ten to twenty years is beginning to make people think of alternatives to flight. People do not usually fly in order to fly; they fly to travel a large distance in a short time. Advances in telecommunications and increases in the cost of travel may reduce the demand for business travel and advances in railway technology will help to reduce medium-range air travel. The ground effect machine, eg hovercraft, may reduce the seaport transfer problem (a good reason for flying over water) by permitting direct communication with inland areas.

The technology of aircraft engines has shown a significant improvement in power to noise ratios and the larger aircraft reduce the total number of flights for a given number of passengers.

All this leads to a certain optimism about the future of aircraft noise but improvements will not happen by themselves. Continued pressure is required on those who have the power to reduce it.

6. Construction, excavation and demolition noise

6.1 INTRODUCTION

The noise due to earth-moving machinery, pile driving and compressor plant operation can reach very high levels. Since many site operations take place in residential areas, sometimes only a few metres away from the nearest house facades, construction noise can be a serious problem. Apart from this, such noise may cause occupational deafness in persons working on site.

It is no longer sufficient to rely upon the temporary nature of site work. In many of our larger cities construction work may be carried out at several points within a small area and for contiguous periods. As a result the annoyance is continuous and merely originates from several different directions over a long period. One-fifth of all complaints about noise concern construction site noise.

Where a large construction scheme is being planned a great deal can be done to influence the attitudes of local residents. A case history of the Neasden Underpass in London, illustrates this point. People were rudely awakened one morning by sheet piling, general construction noise, heavy trucks and neighbourhood disruption without any prior publicity or consultation. Action committees and public protest were swift and vehement. As a result of this great pressure was brought to bear on the contractor and the local authorities. The investigating Public Health Inspector concluded using a medical analogy: '. . . . successful suburban surgery cannot be achieved without adequate informatory service, full public relations and a will to respond to all reasonable requests'.

To quote the British Code of Practice, BS 5228: 'The removal of uncertainty and rumour can help to reduce people's reaction to noise'.

In this chapter we examine the laws relating to construction noise, methods of predicting construction noise and the potential for control at source.

6.2 CONSTRUCTION NOISE LEGISLATION

Different countries have different regulations concerning noise emission. Certain countries allow noise emission up to a certain level during a specific part of the working day and take the character of the noise into consideration; other governments specify the maximum emission outside the nearest building.

In the USA state and local authorities are able to set and enforce limits for construction noise with regard to neighbourhood nuisance under the Noise

71

Control Act (1972). The risk of hearing damage to both construction workers and to similarily exposed members of the public is calculated on the same basis as the Occupational Safety and Health Act Regulations (see Chapter 9). Transportation Authorities, aware of this situation, sometimes devise their own limits. EEC directives adopted in 1975 give member states the task of devising ways and means of meeting emission limits on concrete breakers and regulations covering the whole subject of noise from construction equipment. Noise emission standards for tower cranes, diggers, excavators, and current generators for welding or power supply are likely to be discussed.

Perhaps the most wide-ranging and detailed laws relating to construction noise are to be found in the UK, where one-fifth of all confirmed nuisances are from construction noise. These include The Control of Pollution Act (1974), The Public Health Act (1961), The Health and Safety at Work Act (1974), The Wood Working Machines Regulations (1974), The Land Compensation Act with its associated Noise Insulation Regulations (1973 amended 1975). We look at this legislation in some detail in the following paragraphs.

The Control of Pollution Act (1974) relates to neighbourhood noise and occupational noise. The Statutory Instrument incorporates the Code of Practice for the exposure of employed persons to noise (see Section 9.2).

Under the Control of Pollution Act (1974), local authorities have the power to specify maximum noise levels for construction sites. They can either determine the maximum noise levels on request from the contractor before he commences construction work as part of the consent procedure, or they can impose limits subsequent to the commencement of work.

The Control of Pollution Act contains reference to British Standard 5228 (1975): Code of Practice for Noise Control on Construction Sites. This is an extensive, well-written document and it is highly-recommended reading for anyone concerned with the construction noise problem. It is based on the unit denoted by L_{EQ}. This is the energy equivalent continuous sound pressure level in dB(A) or the steady 'A'-weighted sound level that would give the same total energy at the point considered as the actual varying sound level. The Code recommends that L_{EQ} should be specified over a 12-hour period. The advantages of this choice of unit from the point of view of construction site noise control are that it allows short periods of very high noise levels during longer periods of relative quiet, and that it is very predictable from knowledge of noise-source behaviour. For example

a) If the number of site vehicles past a given point increases by a factor of 10, then L_{EQ} increases by 10 dB(A)
b) If the actual operation only takes up to six hours out of twelve over which L_{EQ} is specified, then L_{EQ} over six hours must be 3 dB(A) less than the L_{EQ} over twelve hours.

The provisions of the Land Compensation Act, Noise Insulation Regulations (1973 amended 1975) discussed in Chapter 4, extend to the noise generated during construction associated with a new highway or a highway-modification scheme. The limit is again specified as 68 dB(A) on the L_{10} index at the affected facade and the noise level has to have increased by at least 1 dB(A) at the point since the construction started.

As a result of this legislation the contractor and the local authority are under the obligation to understand how construction noise propagates around the site and how to choose appropriate equipment to conform with a specified limit. In the following sections we will consider these two aspects in detail.

The individual's rights with regard to annoyance from construction noise remain covered by Common Law.

As a last resort it is always possible for the local authority or the contractor to purchase, ie 'buy-out', the noise-affected households.

6.3 PREDICTION OF CONSTRUCTION SITE NOISE ANNOYANCE

Agreements under the BS 5228 consent procedure could place local authorities in an invidious position if the consent turns out to be mistaken. Nevertheless the consent procedure is the most likely to be invoked. The main problem in predicting the likely annoyance resulting from a predicted L_{EQ} at the noise-affected perimeter of the construction site is that there is little published data. Most of the existing information relates to the BS 4142 procedure (peak levels and background levels in dB(A) – see Chapter 6.4) or to L_{10} index, in dB(A), for road traffic noise. Even if information were available relating annoyance to industrial and road traffic noise measured in L_{EQ}, there is some evidence that construction noise makes people more annoyed than the same L_{EQ} of traffic noise or aircraft noise. L_{EQ} has been used to predict speech interference. For example an L_{EQ} of 65-70 dB(A) implies a sentence intelligibility of only 5 per cent. A United Kingdom Department of the Environment advisory leaflet, 'Noise Control on Construction Sites' (1972), suggested a maximum level of 75 dB(A) at an affected house facade.

In the USA, a Department of Housing and Urban Development Circular (HUD Circular 1390.2 August 1971) established environmental noise standards for new construction sites. If the level exceeds 75 dB(A) for only eight hours in 24 hours it is regarded as unacceptable. This is more stringent than the United Kingdom limit. HUD Circular 1390.2 also suggests that a level which exceeds 65 dB(A) for 8 hours in 24 hours or is "loud and repetitive" is normally unacceptable. The noise measurements are to be made at appropriate heights above the boundaries of a construction site. The builder may not obtain federal funds or guarantees unless the noise levels conform to these standards.

However, assuming that some limit on L_{EQ} is agreed, the next problem is to predict the levels that are likely to result from the construction operation. This is a wise planning move for undertakings in countries other than the United Kingdom also. Since the site might be in a rural area and might result in sound propagation paths of half a mile, one mile or even several miles in length, prediction is not always straightforward.

In the same way as the intensity of thermal energy from a fire grows less with distance and with shielding, so does sound energy from a piece of construction site machinery. For short distances from the equipment it is possible to assume a decrease of 3 dB per doubling of distance in L_{EQ}. However for large distances the attenuation depends principally upon the height of the propagation path. Topographical features of the site can have a considerable influence in the attenuation over long distances and should be used thoughtfully during construction. Indeed BS 5228 suggests the use of spoil from site excavation to construct an earth bank between the construction equipment and nearby housing. Otherwise the site should employ specially constructed barriers. If conditions on site are at all unusual the current state of the art requires that on-site propagation trials be carried out. However BS 5228 points out that at present there is no great experience of the ground attenuation to be expected at construction and demolition sites and for the time being it is suggested that this factor be ignored.

6.4 CONTROL OF CONSTRUCTION NOISE

It should be remembered that use of L_{EQ} implies a trade-off of level versus time. In other words, very noisy machinery is not prohibited from use as long as its time of operation is carefully controlled. BS 5228 gives advice on times of working and recommendations for the reduction of permitted levels outside normal working hours. Selective use of equipment and control of noise propagation therefore remains a viable alternative to modification and/or replacement of equipment. We should not forget, of course, use of alternative construction schemes, techniques or locations; however, these options may be restricted by cost considerations.

The sort of questions that should be asked concerning location are: where is the most suitable location for compressors? where is the most suitable place for vehicles to enter or leave the site? at which part of the building site should work begin in order to minimise the strain on people living near by? For example, in certain cases compressors are least disturbing when located in the middle of a building site. As the excavation work on the building site proceeds they can be moved down into the excavated areas where the noise can be directed into the building site or upwards away from sensitive areas. As a result of the evidence on potential noise and other disturbances delivered at a public inquiry in the United Kingdom where a Fullers-earth mining firm were appealing against a local authority rejection of a proposed open-cast excavation site, the inspector ruled that the mining operation could be allowed if the firm reversed their proposed plan of excavation and started at a point furthest from the nearest residences. He also suggested that the firm sacrificed what, according to the proposed plan, would have been their first year's operation at the highest point on the site. This sacrifice, together with the dumping of spoil, would form a barrier against noise nuisance. Such steps can thus permit the exploitation of a site while minimising disturbance.

Other factors can be borne in mind when operating a site. For example, a compressor which is sited near a wall will produce 3 dB(A) more than one which is positioned well away from any vertical sound-reflecting surface. If positioned in a corner where two walls meet, there may be an addition of a further 3 decibels. Such simple points can save the undoing of any advantage achieved by using the quietest possible equipment.

Another point concerns the decision on the number of compressors. For example, there may be a choice of compressors with an output ranging from 3.5m³/minute up to 20m³/minute, each with a noise emission level of 70 dB(A) measured at the nearest building. Suppose that 38m³/ minute of air is needed on a building site. One choice is to use four compressors each delivering $9.5m^3$/minute. The noise level emitted by each is 70 dB(A). All four when lined up together will emit a combined 76 dB(A). A further choice would be to use only two compressors, each delivering $20m^3$/minute. These two compressors combined will emit 73 dB(A) measured at 7m. It is preferable therefore to use a few large compressors in place of several small ones when the sole purpose is to reduce noise.

For most construction sites in urban areas, with the exception of pile-driving or impact equipment of some kind, a basic noise level of about 80 dB(A) L_{EQ} at the site perimeter is set by the operation of trucks during the period of their operation. It is possible to consider the noise control options against the associated costs so that a realistic choice can be made. A summary of the total L_{EQ} for each possible construction method and noise control strategy can then be drawn up. Among the strategies that could be considered are:

1) variation in type of construction scheme,
2) application of 'state of the art' noise control to the *noisiest* equipment in each scheme,
3) application of *advanced* noise control to the noisiest equipment in each scheme,
4) application of 'state of the art' noise control to *all* equipment in each scheme,
5) application of *advanced* noise control to *all* equipment in each scheme,
6) specially designed equipment (pile driver rigs such as 'Taywood' or 'Hush') for use in place of extremely noisy machinery.

Each figure should relate to the estimated equivalent level over the whole of the construction activity. State of the art techniques are those already available in the form of noise reduced equipment (see details in BS 5228 and Table 6.1). The implication should be clear, that use of 'quiet' equipment is far more

Table 6.1   Typical peak sound levels from construction equipment; silenced and unsilenced

Equipment	Sound level at 7m in dB(A)
Unsilenced pile-driver (drop hammer type)	110
Unsilenced truck, scraper or grader	94
Unsilenced pneumatic drill	90
Unsilenced compressor (hydraulically operated) Concrete breaker Muted pneumatic drill Cranes (movable and derrick)	85
Sound reduced jack hammers and rock drills Drop hammer pile driver fitted with acoustic enclosure Unsilenced generator Truck with efficient silencing	82
Muted pneumatic drill fitted with sound-damped steel Electrically operated drill	80
Silenced crane	75
Sound reduced compressor	70

75

noise-effective than variations in construction strategy in terms of impact on L_{EQ}.

The European Economic Commission has issued a directive on the measurement of sound power levels from construction equipment and has laid down emission standards for road-breakers.

With regard to reduction at source, the provision of a light exhaust muffler effectively reduces the operating noise of a pneumatic drill or concrete breaker without loss of performance. The fitting of sound deadening steel produces a further reduction. Enclosing the hammer head and top of the pile in an acoustic screen reduces the operating noise of a pile driver considerably. On a sound-reduced air compressor the air cooling system is designed to enable the compressor to operate with a canopy fully closed on all sides. The canopy and chassis are lined with sound-absorbing material. Noise from prime movers, particularly exhaust noise because of age or damage, is often the most offensive. This applies to compressors, compressed air equipment, welders, mobile cranes and earth-moving equipment. On all of these an improvement of 5 – 10 dB(A) is possible by improving the silencer or muffler. Engine noise is more difficult to tackle since cladding can lead to overheating. Moreover, electrically-driven equipment, although quieter than diesel powered devices, can be risky in wet weather.

An alternative to the conventional noisy impact type of concrete breaker breaks concrete by bending – a process that produces no more noise than that of the site vehicle to which it is attached.

The quieter future of construction noise from pile-driving probably lies with improved manufacturing specifications and with the application of damping to piles or with the development of new techniques for piling.

6.5 OPEN CAST MINING AND EXCAVATION

A recent ruling by the Department of the Environment in the UK suggests that BS 4142, 'Method of rating industrial noise affecting mixed residential and industrial areas', is an appropriate basis for assessing the likelihood of annoyance from existing workings and for forecasting this likelihood for projected workings. This standard is described in the next chapter. Where forecasts or predictions are to be made, it will be important to establish the exact types, numbers and sound power levels of the excavation plant and the method of working the site. This information can then be coupled with the prediction method for each individual plant as outlined in BS 5228 to establish the likely peak levels during working.

6.6 HOW TO DEAL WITH CONSTRUCTION SITE NOISE

Whether or not construction site noise is of a temporary nature, it does cause quite an unusual number of noise complaints. Very often these complaints arise because people focus on the noise of the construction process itself as a tangible thing to complain about, whereas what they are really disturbed about is the change in their environment.

There is relevant legislation in most countries. In the UK, the normal process of Common Law and the action of the Control of Pollution Act (1974) can be used to deal with this sort of noise. Often the defence is advocated that the noise is of temporary nature and that the complainant has only to be patient for it to go away. The noise that will occur following the construction of new works, in particular a motorway, is a separate issue which in the United Kingdom is dealt with under the Land Compensation Act 1973, amended 1975 (see Section 4.6).

Persons intending to deal with construction site noise in the UK should consult the Code of Practice BS 5228, which deals with the amount of noise that can be produced during the process of construction. This Code of Practice quotes probable amounts of noise that various items of construction machinery will produce.

Local authorities should be pressed to set limits to noise levels and also limit the times of day when particularly noisy operations, such as pile driving, can take place, and contractors should be pressured to use quieter equipment in construction work. This should normally be done via the local authority. Construction workers are at risk to hearing loss due to construction noise. In almost any large city one can find persons operating pneumatic drills. Any such person is probably going deaf at quite an alarming rate since it is unusual to see any of these people wearing ear defenders. Although some effort has been made by the manufacturer of road drills to reduce the noise from the drills themselves, usually this effort is confined to an attempt to reduce the noise from the exhaust of the compressed air which drives the drills. The main source of the noise – that is, the resonance of the drill bit itself – is usually left untreated, necessarily so, as it could not do its job if it were encased in sound-deadening barriers. It is urgent that it be made mandatory that all persons operating noisy construction and demolition equipment should wear ear defenders. Pressure can be put on building trades unions to see that they are aware of these facts. There is quite a lot of psychological resistance to this as the operators themselves feel that it is unmanly to admit that they need protection from such an intangible thing as noise. However, it has been shown that it is possible to overcome such prejudices. The precedent here is the use of the hard hat. Once it had been established that the wearing of a hard hat was a status symbol showing that the wearer was engaged in a hazardous and therefore somehow prestigious operation, the hard hat became very acceptable to workers. In order to reinforce this process the wearing of the hard hat was made mandatory for all persons engaged on the construction work and all persons of authority who came to investigate the site were also made to wear hard hats. There were no exceptions. Pressure groups and local authorities should strive to see that such a similar scheme be implemented where loud noises are found.

Audiometric testing and screening of workers in the construction industry is as essential as it is for those working under high noise conditions within factories (see Chapter 9).

The Code of Practice for noise control on construction and demolition sites (BS 5228) gives guidance on current thinking. Under Section 61 of The Control of Pollution Act (1974), a person wishing to invite tenders for a construction or demolition contract shall be entitled to notify the local authority and to require them, within a reasonable period, to serve a statutory notice upon the contractor specifying the requirements to be observed in the execution of works for the prevention and mitigation of nuisance from noise or vibration. Also, local authorities can serve notices in respect of works of which they have not had formal notification under Section 60 of the Act. There is provision for prompt appeal to a magistrates court against the requirement of a notice and in appeal proceedings it would be open to the appellant to show that the requirements were sought to be in excess of the general duty. Even after such an order had been served however, it is possible to act against demolition and construction companies under common law and Section 59 – requirements that they would normally have to comply with. Also, if the contractor fails to comply with the requirements of a notice the local authority is able to apply to a Justice of the Peace who would be able, if he thought it right after inspecting the works, to order their suspension until such time as the contractor satisfies a magistrates court that he is able and willing to comply with the requirements. This means that work can actually be stopped if excessive noise is being caused.

6.7 CONCLUSION

The problem of construction site noise has only recently been brought under active attack in several countries. It is too early to assess the likely effect of the legislation. However, the legislation does appear to be reasonable and, it is hoped, will be effective both in causing contractors to reduce the noise from activities and to specify quieter machinery. Such quieter machinery is already available because of stricter European noise requirements, particularly in West Germany, and so it will not be difficult technically to reduce noise in this context. Nevertheless, it is important that individuals and local authorities maintain the pressure on construction companies to see that what is technically possible actually happens.

7. Neighbourhood noise and noise due to industrial premises

7.1 INTRODUCTION

Patterns of human settlement are becoming more dense. The predominant choice of human settlement is urban living. Studies show that approximately eighty per cent of the US population are living in urban areas. This figure is exceeded already in the United Kingdom and several European countries. A by-product of this trend is an increase in neighbourhood noise and the impact of industrial noise on residential communities. In contrast to six out of nine of the countries in the European Economic Community, the USA leaves the major role in controlling neighbourhood noise to the state and local governments, retaining only the right, through the Environmental Protection Agency (EPA), to establish national uniform noise standards for major products and sources involved in interstate commerce. In fact a compilation has been made of the wide variety of municipal noise ordinances, which provide the principal form of control, and the EPA has provided a model community noise control ordinance. The municipal ordinances offer a wide range of permissible levels, usually specified in octave bands measured at the property line of an individual residence or at the boundary between a manufacturing and residential district. Some ordinances differentiate between limits for day (7 am to 10 pm) and night (10 pm to 7 am); some assume that limits that are acceptable for daytime are applicable also to the night-time period. Many ordinances include an adjustment for pure tone components. Different levels are sometimes specified for each of the following types of area: heavy manufacturing zones; commercial zones and residential zones. The local authorities may require compliance with these levels if exceeded within 12 – 18 months. Additional controls may take the form of restrictive zoning whereby certain forms of land use are not permitted to occur on the borders of a residential area. Much of this is similar to, if not as detailed as, the relevant local authority action in the UK discussed later.

Of the EEC countries, Germany has set specific targets, in its 1968 technical guidelines for noise control, of a maximum of 50 dB(A) for the L_{EQ} during the day and 30 dB(A) for the L_{EQ} during the night. Measurements have to be made 3 m from the boundary and 1.2 m above the ground.

Until recently in the UK the reduction of noise around factories has occurred, if at all, as a result of specific complaints from neighbours. The factory makes the noise, nearby residents complain and sometimes with the help of a local authority force the factory management to take some noise-abating action. Now however, in the UK a new concept has been introduced – the concept of the Noise Abatement Zone (NAZ). Under the noise provisions of the 1974 Control of Pollution Act it can be made compulsory for the factory management to keep the noise from their factory below a certain level at the boundary of their premises if the factory is within a noise abatement zone. So it will not be necessary to fight individual actions to reduce factory noise within the NAZs. Outside the zones, however, different legislation still applies and so we must also consider how this works.

7.2 COMMUNITY NOISE LAW

Details of the law on community noise are given by Kerse and the Noise Abatement Society (see Bibliography). Under UK statutory law a noise nuisance becomes a statutory nuisance when it is declared a nuisance by a magistrate's court or by a local environmental health officer. Often, though, the best practicable means defence is used to show that although the noise is high it cannot be reduced by economic practicable means and that the best measures, bearing in mind cost factors, have already been taken.

7.3 COMMON LAW

In addition to the statutory law there is a wide body of common (case) law which has been built up over many years and which can be used to protect the owner of property against nuisances, including noise. In this case in the UK, a nuisance is described as 'An inconvenience materially interfering with the ordinary comfort, physically of human existence not merely according to elegant or dainty modes of living but according to plain sober and simple notions among the English people'. A property owner can bring an action in court for noise nuisance. The advantage of this form of action is that the best practicable means defence is not available to the noisemaker.

7.4 BRITISH STANDARD 4142 'METHOD OF RATING INDUSTRIAL NOISE AFFECTING MIXED RESIDENTIAL AND INDUSTRIAL AREAS'.

BS4142 is a document that is meant to advise local environmental health officers whether a noise from a fixed industrial process on industrial premises is unduly noisy. The standard is used primarily when a member of the public has complained about being subjected to a noise nuisance. During the investigation and any subsequent legal action involving the service of a statutory notice on the offender, regard is paid to the physical characteristics of the noise and the method of assessment of its likely annoyance is provided by the procedure set out in BS4142.

The procedure laid down is reasonably straightforward and with the aid of the revised document BS 4142 (1974) and a sound level meter any intelligent person can form an opinion about the nuisance caused by such a noise.

However local environmental health officers are sometimes jealous of their positions and will state quite rightly that they are free to use their own judgement in particular cases. So care must be taken when using BS 4142 conclusions at a planning enquiry or appeal. Care must also be used to see that BS 4142 does apply to the case in point. However, in the absence of any guide to what is an acceptable level of noise, such as would face the occasional user of a helicopter at a factory, BS 4142 can be carefully cited as a guide to likely noise nuisance.

Essentially, the standard lays down a maximum possible level difference between the offending plant, of which the measured level is adjusted according to certain characteristics of the noise, and the ambient or background noise (this is either measured or estimated where measurement is difficult). The maximum permitted difference is 10 dB(A). However if this is exceeded it does not automatically mean that the plant owner has broken the law. All that the standard means to the person using it is that complaints are very likely when the criterion (background noise level) is exceeded by 10 dB(A). In such cases where the difference is between 5 and 10 dB(A) there is a borderline possibility of nuisance. When the difference is zero or less, there is no justifiable case of nuisance. In practice, any environmental health officer would use the standard merely to give himself an indication to justify his conclusions about the need for nuisance abatement.

Creeping noise nuisance
As a result of advice given in Circular 10/73 'Planning and Noise', local environmental health officers often face a difficult situation when dealing with new factories or 'improvements' to existing factories. The standard reference BS 4142 apparently states that no justifiable noise complaint will occur if a factory makes a noise up to 5 dB(A) *above* the existing noise level. If this is allowed to occur however, the new higher noise level can become the new background noise level for yet more noisy activities which, the new noise makers will argue, can be yet another 5 dB(A) higher. This process can be repeated and leads to creeping noise nuisance. Against this, the local environmental health officer has 'state of the philosophy' arguments (the spirit of the newer legislation is against creeping noise nuisance) and recourse to the involved procedures of a noise abatement zone. However, where alterations require planning consents a noise clause can be added as a condition of the consent. This is a powerful weapon in the hands of the local authority. It should, however, be used carefully since any overstating of nuisance potential or prescription of unusually low limits can be overruled by Department of the Environment inspectors at planning appeal stage.

7.5 THE CONTROL OF POLLUTION ACT

This Act makes it the duty of every local authority to detect any noise nuisance that ought to be dealt with by them and to decide how to use its powers concerning Noise Abatement Zones. This does not mean that local authorities have to create these zones but that they must in a general way continue to keep an ear on local noise problems. The Act also lays down rules for noise from construction sites (described in Chapter 6).

The main noise provisions of this Act are contained in sections 58 to 61. Section 58 deals with the power of local authorities to deal with noise nuisance. Section 59 deals with the action to be taken by individuals against noise. Section 60 deals with the action to be taken by local authorities against construction site noise. Section 61 deals with the prior consent that construction companies can obtain to make noise before starting to work.

7.6 NOISE-ABATEMENT ZONES

A local authority can decide that any or all premises, except domestic premises, that fall within its area are within a noise-abatement zone. When this decision is made, however, the authority must take certain actions.

The local authority must issue an order, to be confirmed by the Secretary of State for the Environment, specifying the area of the NAZ and listing the classes of premises within the area which are affected. This order is called a "noise-abatement order". All owners and occupiers of scheduled premises within the NAZ must be informed and the order must be confirmed by the Secretary of State after appeals and enquiries have been heard. The authority must then measure the levels of noise at points along the boundaries of the relevant premises within the NAZ. These can include industrial and commercial premises, places of entertainment, transport installations and public utility and agricultural installations. It is required that the local authority prepare a 'noise level register' of the measurements. A copy of the entry in the register must be sent to the owner and the occupier of the premises concerned who may appeal to the Secretary of State within 28 days. It is up to the local authority to decide which classes of premises are to be included and when to make the measurements. The public should be able to inspect the noise level register free of charge during reasonable hours.

Once a level has been entered in the noise register it cannot be exceeded except with the written consent of the local authority. It might appear from this that the owners of premises should hurry to install noisy plant before the measurements are made and to ensure that such noisy plant is operating when the measurements are made. However, if it appears to the local authority that too much noise is emanating unnecessarily from any premises, before entering any levels into the register the authority can issue a 'noise reduction notice' on the premises requiring a reduction of noise within a period of time not less than six months. The notice can be specific about which noises are to be reduced, by how much, and when (if at all) the noise can be permitted. The owners or occupiers can appeal to a magistrates court against the notice.

Details of special permission to exceed the registered noise level or of noise reduction notices should appear in the noise level register.

7.7 LAND COMPENSATION ACT 1973 AND INDUSTRIAL NOISE

In theory the Land Compensation Act (1973 amended 1975), although used mainly to deal with the problem of new and altered highways as far as noise is concerned, should be appropriate to deal with nuisances arising from the altered use of neighbouring land in connection with a wide set of statutory purposes.

The road traffic noise aspects are dealt with fully in Chapter 4 and it is unfortunate that there is as yet no statutory guide under the Land Compensation Act equivalent to the 'Noise Insulation Regulations' to deal with compensation claims arising from noise from sources other than road traffic.

In fact the act is couched in vague terms and except for the road traffic case appears to be of little concrete use at all.

Nevertheless a resourceful landowner backed up with measurements, BS 4142 assessments, and a good legal counsel might be able to establish that a compensatable nuisance has been inflicted upon him by alterations in the use of adjoining premises and will then probably be able to gain compensation or sound insulation similar to that described in the Noise Insulation Regulations.

However, the authors know of no such case, successful or otherwise. For the moment anyone trying this field will be pioneering and had better be well-equipped or, better still, should wait for further regulations to be issued or for a well-set-out precedent.

7.8 DISCUSSION OF THE GENERAL SITUATION RE INDUSTRIAL NOISE

We can see that there are several ways, legal and advisory, of establishing whether an industrial noise is a nuisance – that is if the complainant is justified in complaining, but both the law and the local authority have to balance the needs of industry to operate and thus provide jobs, products and exports for the common good on the one hand, and the need of the individual for a quiet environment on the other. This means that not all complaints about noise will succeed and in the event of failure a complainant must find his own remedy and this might mean a move away from the noise.

Individuals or factory managements are in a still more difficult situation when it comes to decisions over the siting of new industrial plant. The Noise Abatement Zone procedures, where applicable, move the responsibility from the shoulders of themselves on to the local authority who, by specifying levels, could effectively exclude certain industries from any zone. However, whether NAZs are in force or not, the main difficulty is that of predicting levels at the boundary of the premises from the plant to be installed. We touched on some aspects of this in Chapter 6. The industry or an individual amenity group may wish to appeal against a local authority decision about the siting of a new factory. In which case there is likely to be a planning enquiry where the cost/benefit balance has to be made by the ministry-appointed inspectors. We explore cost benefit analysis as it relates to noise problems in Chapter 10.

7.9 PROCEDURES FOR HOUSEHOLDERS SUFFERING EXTERNAL NOISE

If the noise is an industrial one emitted from industrial premises, initially the householder should lodge a complaint with the local Environmental Health Officer, who is then obliged to investigate the complaint. He will probably proceed using the methods laid down in BS 4142 or their local equivalent and need not necessarily agree that the noise complained of is a nuisance. If the local authority in the person of the local Environmental Health Officer thinks that the noise is a nuisance, he will serve an abatement notice on the offender. The abatement notice will require the abatement of the nuisance or the restriction or prohibition of its recurrence.

If, however, the local authority thinks that the noise is not a nuisance, the Environmental Health Officer will inform the complainant of this fact and is not obliged to take any further action.

Sometimes the local authority will have licensed a landuser to carry on a particular activity, typically an outdoor sport, and could and should have included noise limits as a condition of the granting of the licence. This can be a powerful means of noise control and should be used wherever possible by local authorities.

7.10 APPEALS AGAINST ABATEMENT NOTICES AND OTHER CONSEQUENCES

In the United Kingdom, the person receiving a noise abatement notice can appeal to a magistrates court within 21 days from the receipt of such a notice. This effectively means that he can go on making the noise until the appeal is heard so that temporary noises cannot be dealt with in this way. Even when an appeal is denied or is not made, it is not certain that the noise will be abated. A factory owner may ignore the order and pay any fines incurred as a business expense (cynically deducting them from his taxable profits); however the fines could be substantial. It is also possible for a notice to specify an immediate halt to the noise and the Environmental Health Officer must specify why this has been done. Care must be taken, however, since if in subsequent action the abatement order was found to be unjustified, the local authority could be liable for damages representing losses incurred by the noisemaker as a result of being forced to stop the noisy process. This form of instant abatement notice is most commonly used to stop noisy parties in crowded urban areas.

More legitimately, the producer of a noise may plead the defence of the 'best practicable means' – that is, he will try to show that he has taken all possible steps to reduce the noise and that further action is impossible (or at least uneconomical). Under these circumstances the noise will continue. The defence is only available to businesses.

7.11 DEALING WITH NOISE INSIDE A NOISE ABATEMENT ZONE

The householder or other complainants should determine whether the industrial premises producing the noise are or should be within a Noise Abatement Zone. A Noise Abatement Zone can include any premises except premises that are used for residential purposes. It should, but need not necessarily, include all the industrial noise sources within the area and at least the properties mainly affected by noise from any of them. The zone is unlikely to include any property significantly affected by noise from a source outside the zone. Under these circumstances the householder could argue to the local authority that the Noise Abatement Zone should be extended to include the external source.

Having established that the noise is within an NAZ the complainant should consult the noise register to see what level the premises are permitted to produce at their boundaries. They should check that the actual levels are below the permitted level. They can do this privately by engaging a noise consultant or by pressing the local Environmental Health Officer to check the levels himself. If the levels are above the permitted levels, a complaint should be lodged with the local Environmental Health Officer who ought then to serve an abatement order. If he does not agree that the noise is excessive, he may well give permission to exceed the registered level to the noise maker who can then carry on as before.

7.12 ACTION BY INDIVIDUALS

Having exhausted the possibility of action via the local authority, a noise sufferer must then take action through common law (see Section 7.3). This is only available to landowners and will involve legal proceedings which may be expensive, although they can lead to an injunction and damages. Another means of procedure is for an individual to apply to a magistrate for a nuisance order, under Section 59 of the Control of Pollution Act (1974). Such an order is similar to the order obtainable by the local authority if they think fit. If the application is successful, the noise will be treated as if the local authority had issued an abatement notice (see above).

7.13 ASSESSING A NOISE NUISANCE

An industrial noise can be assessed using the procedure of BS 4142 or by checking measured levels along the boundary of the industrial premises against registered noise levels where the industry falls within a NAZ. For either check a precision sound level meter is required and the services of a local noise consultant are usually needed. A local university or polytechnic will normally contain a department from which one of these devices and measurements (involving, say, half a day's work) can usually be obtained for a fee. It is a mistake to think that such services should be free. College staff have their own work to do and will usually have to come to some arrangement with their college authorities before such private work using college equipment can be undertaken. Alternatively, an independent consultant can be employed.

Having obtained a measure of the loudness of the sound in dB(A) using sound level meters, other factors must be taken into account. These are: type of noise, frequency and duration of the noise, character of the district, newness of the noise and the tonal or impulsive nature of the noise. These factors are taken into account by adding or subtracting corrections to the measured noise level. This measured level is then compared with either the actual background level* measured without the intruding noise or else an estimated or notional criterion calculated from the remaining factors listed above. If the corrected level exceeds the corrected criterion by more than 5 dB(A) the noise may be regarded as a nuisance. In practice, it would be unwise to proceed unless this level difference was at least 5 dB(A) and even then there is no guarantee of satisfaction because the noise is not a statutory nuisance unless declared so either by a local authority, who may choose to ignore protests, or else by a magistrate if an individual can make a case.

7.14 LATER STAGES OF NOISE ABATEMENT ACTION

It is not clear what occurs in the later stages of Noise Abatement Action since most disputes are settled out of court to the satisfaction of both parties or else the offending party proves that there is no case to answer and the situation continues as before. If this happens the only remedial action available to the noise sufferer is to move away from the noisy situation, and he should do so. Often people dig their heels in, determined to abate the noise nuisance when it is patently obvious that they cannot do so. Such people become 'noise

---

* This is defined as L_{90} by the 1975 amendment to BS 4142. For methods of compiling L_{90} by spot readings using a sound level meter see Chapters 2 and 4.

hobbyists' and it can be said that they are not really upset by the noise. Often they are, in fact, enjoying it because it brings them a chance to make a nuisance of themselves to the local authority, produce action groups and generally get themselves known in the community. This can be justified in the case of a major noise nuisance which will affect wide areas and large numbers of people, such as a new airport, but for small domestic disputes involving a handful of people most of whom are not particularly bothered by the noise anyway, this action can only be regarded as irrational. If it is remembered that only 20 per cent of the population appears to be unduly upset by noise, it is to be expected that only one in five of the neighbours will agree that any particular noise is a nuisance.

7.15 THE BEST PRACTICABLE MEANS DEFENCE (AS USED IN THE UNITED KINGDOM)

This defence is open to any business which can show that they have used the best practicable means to reduce the noise and that it is impracticable to reduce it further. Factors taken into consideration include cost as well as physical factors. It is almost always possible in theory to reduce a noise, but more often than not such a reduction can only be achieved at great expense. The operation of the defence will probably involve some scrutiny of the books of the business involved, and that might deter some businesses from using it. The defence is available against actions under sections 58 and 59 of the Control of Pollution Act (1974) and also against reduction notices served under the Noise Abatement Zone procedure. It is not available against owners or occupiers of land who bring a common law suit of nuisance (see Section 7.12).

7.16 CONCLUSION

The improvement in means to deal with neighbourhood noise in recent years has been quite substantial. Not all noises can be dealt with. Some noises must be tolerated because of the benefits arising out of the noisy process. Sometimes it is better for the sufferer to move if he is alone in suffering and if a successful legal action would cause a valuable source of employment to be closed down.

8. Domestic noise problems and the noisy neighbour

8.1 INTRODUCTION

The noise from neighbours is an insidious thing in that it cannot be dealt with at a distance. So often this type of noise problem can involve face to face contact and domestic strife. The result of this is that many people will tolerate unnecessary noise from neighbours either from a desire to avoid trouble or, perhaps more honestly, in the knowledge that the neighbours must tolerate their own noise in return. As a result, many people will get away with making a noise nuisance of themselves without ever being aware of the fact. Noise from neighbours is caused by TV, radio and other hi-fi devices, cars warming up, burglar alarms, footsteps overhead, shouting, musical instruments, children and dogs. Noise from within dwellings is caused by fans, pumps, fluorescent lamps, boilers, pipe work, lifts and doors.

8.2 THE NOISY NEIGHBOUR

The noisy neighbour is best dealt with at a personal level. A visit or a telephone call or a letter politely explaining a grievance can sometimes help. Writing anonymously or as a group of neighbours can overcome problems of shyness or embarrassment or fear. However, anonymous letters are dangerous and should be in no way abusive or obscene or the police will take an interest.

Often a neighbour will discuss the point and will make less noise (never insist on no noise) provided he can do so without loss of face. Frequently, however, there will be an initial adverse reaction. For example, the chronic car-warmer-upper-at-7.00-a.m. will make twice as much noise after being spoken to but this is usually a temporary phase and his or his family's desire to be accepted in his neighbourhood will win in the end. Do not react to temporary provocations of this sort.

Abusive shouting or noisy parties can be dealt with by a call to the police. They are usually quick to respond to such calls and although there is little they can do directly, a resourceful policeman can find surprisingly effective ways of 'leaning' on a noisy group of revellers. Abusive shouting can be a symptom of mental ill-health and the sufferer – that is, the shouter himself – can have a doctor sent to visit him by the police and will usually co-operate by taking suitable drugs if undesirable alternatives are pointed out to him.

The typical response to annoyance from a neighbour's radio, hi-fi or TV is to return the annoyance in kind. A discussion or mutual agreement is to be preferred since the resulting noise can be offensive to those not party to the original dispute. Often, however, especially in apartments, the annoyance can result from poor sound insulation between adjoining dwellings.

Action on sound insulation
It is practically impossible to improve substantially the sound insulation of an existing party wall that does not have an obvious defect. When moving into a multi-occupancy dwelling or a semi-detached house, it is prudent to assess the neighbours' life-style and hi-fi resources and TV channel preference (try and get in for coffee). Their children and their ages and any dogs need consideration. Conversely, consider the likely impact of your noise profile on them. Don't move in next door to anyone who seems noise-sensitive if you are given to noisiness yourself.

If you are bothered by noisy neighbours there is only one really effective solution and that is to move. If you consider yourself to be particularly sensitive to neighbour noise, do try to obtain a detached dwelling. If you are housed by the local authority, money can be difficult and you may have to trade one disadvantage (the noise) for another (an inconvenient location). It is also necessary to be truthful with the housing office: it is necessary to tread a narrow path between making your noisy neighbour appear to be intolerable (how are they going to replace you?) on the one hand and on the other hand, you should not appear to them to be over sensitive ('this person will never be satisfied'). Try to bargain for a once and for all move from a problem that affects you personally.

Sound insulation between dwellings is largely determined by the mass per

Figure 8.1   Relationship of sound reduction to superficial weight

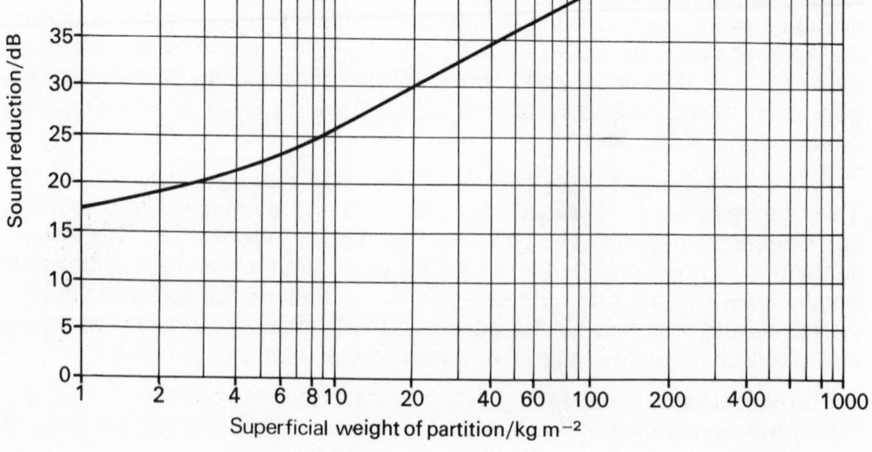

unit area of the party wall (Figure 8.1). Small defects in the wall, such as cracks or holes can dramatically reduce the insulation but these are rare. Often the sound insulation of a wall can be adversely affected by flanking transmission. Sometimes the sound travels round the edge of the wall because windows are placed too close to the party wall on both sides. In older dwellings the party wall is sometimes not carried up to the roof, leaving a flanking path between the two adjacent thin plaster ceilings.

In the UK the Building Regulations (part G, 1965 amended 1972) specify the sound insulation to be provided between dwellings but these regulations are hard to enforce and a dwelling only has to meet the regulations in force at the time of construction. There are also *deemed to satisfy* clauses in the regulations which allow party walls or floors of specified construction to be acceptable whether or not their measured insulation meets the standard. In practice the noise provisions of the Building Regulations are often only to be adhered to when a client, usually a local authority, has specified that they shall be adhered to. It is also doubtful if the standard provided by the Building Regulations is adequate now that houses no longer have large chimney breasts taking up a high percentage of the party wall. Conversely, it has been argued that the low frequency insulation requirement (40 dB at 100 Hz) is excessive, particularly since low frequency insulation is difficult to achieve with light-weight double panels. It is also true that building regulations on a federal basis do not exist in the USA.

Sound insulation treatment, if all else fails, is largely a waste of money and effort. Indeed some forms of so-called insulation can actually increase the amount of sound passing through a wall. Thin plasterboard on battens can often do this. The plasterboard resonates against the air in the cavity and this resonance increases the amount of sound transmitted. Filling the cavity with fibreglass helps, but the Mass Law (Figure 3.10) is basic against all such treatments. The Mass Law means that to achieve a 5 dB increase in overall sound insulation, the total mass per unit area of the wall must be doubled (and doubled again for a 10 dB increase). Given a situation where the mass of the wall is already substantial and an improvement of at least 10 or 20 dB is needed for satisfaction to be achieved, the impossibility of the task becomes clear.

8.3 CENTRAL HEATING AND AIR CONDITIONING SYSTEMS

Noisy thermal control systems are a common cause of nuisance, especially in blocks of flats and in dwellings near refrigeration plants such as are often found outside grocery stores and butchers shops. Although this is not common in the UK, it can happen that neighbours operate a fan or air-conditioning unit so as to cause a noise problem.

Quite often, builders install units in good faith because manufacturers' estimates or measurements of the noise levels from the devices appear to be quite low. It is a great mistake to rely on such information alone, since it is very often the vibration from the device which causes the noise problem and not the airborne sound. Thus when, for example, a fan is mounted on a roof over a ventilation duct leading up from a bathroom or kitchen, the entire roof and thus the ceiling of the room space below can be forced into vibration and act as a

Figure 8.2 Treatment of airconditioning fans

Louvered intake with sound absorber

Barrier near intake

Service connection with flexible section

Massive base

Antivibration mounting

Radial fan

Canvas bellows

Sound absorbing section with splitter

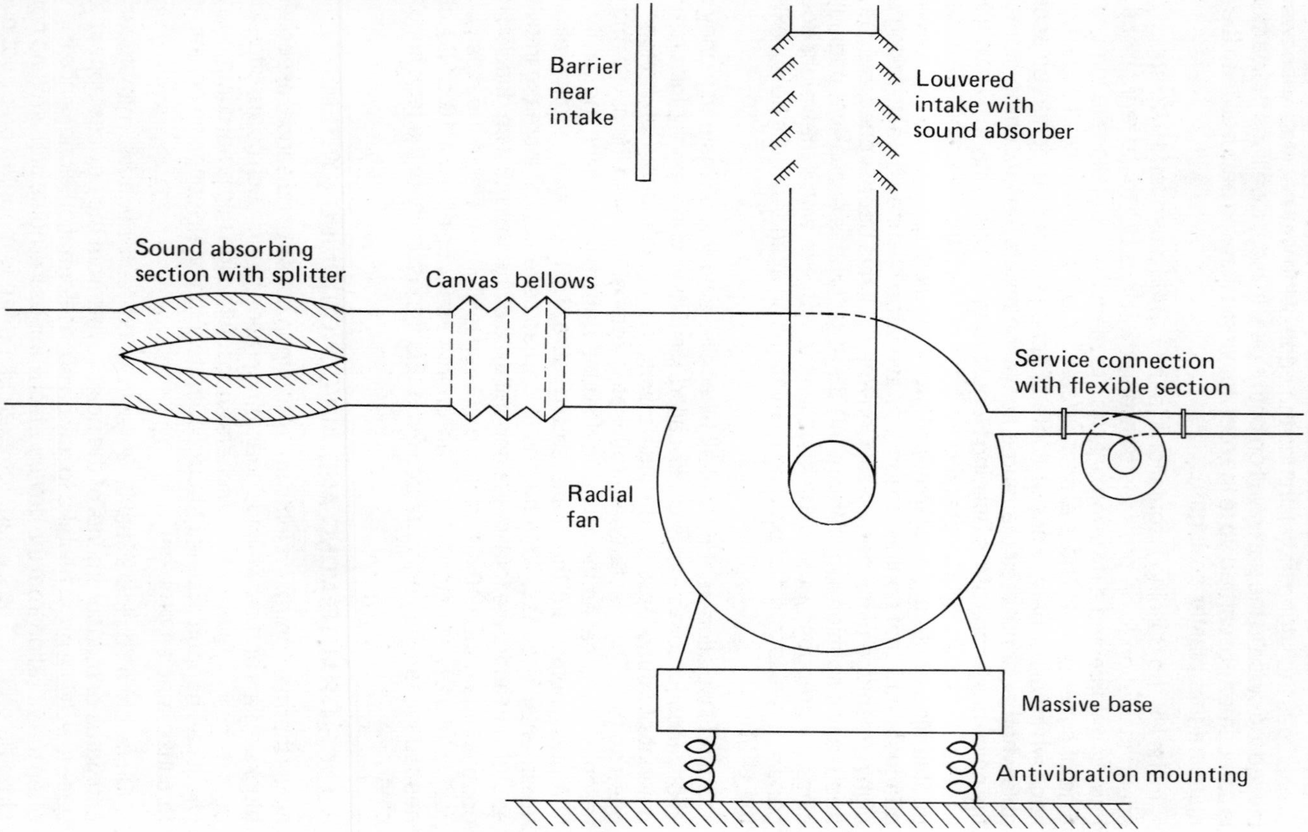

loudspeaker producing quite unacceptable noise levels. This problem can be cured by properly designed anti-vibration mountings. These usually consist of a flexible bellows which de-couples the fan from the rest of the ducting and coil spring anti-vibration mountings to support the fan itself. Such coil springs should be strong enough to support the fan without being over compressed, but must have enough 'give' to allow the fan unit to settle a short but noticeable distance, about 10 mm or more, when installed on the spring. Such treatment will often suffice to reduce considerably the noise from the fan. For more severe cases it is sometimes necessary to treat the ventilating system completely. Such treatment is shown in Figure 8.2 and consists of anti-vibration breaks, airborne sound attenuators and anti-vibration supports.

Ventilating systems can also cause noise by being overloaded. If the air is forced down the duct at too high a speed or pressure, turbulence can result from friction at the duct walls or at pressure-reducing devices. In less extreme cases, badly-designed exit grills over a ventilation point can also cause turbulence and hence noise. Such grills should be replaced by more openly-designed ventilation ports and while this is carried out a check should be made by listening for noise travelling down the duct from other trouble spots. It is not true that sound is blown back by an air current, since the speed of sound is some 350 m/s and this far exceeds the air speed in normal domestic ventilation systems.

8.4   FAN NOISE – WHAT TO DO

This section is intended as a procedural example of how to tackle a specific noise problem.

1. Investigate the fan.

   a) Is it bolted directly on to the airduct(s)?
   A fan which is directly connected to an airduct can send vibrations down the duct wall. These can get out of the structure at any point of support along the path.

   b) Is the fan bolted directly to the wall or roof?
   A wall or roof can be excited into high vibration levels by a fan and this can cause intense noise in the rooms below, where the ceilings can behave like huge loudspeakers.

   c) Are there any anti-vibration mountings?
   The easiest way to test for anti-vibration mountings is to push or lift the fan gently. If the fan gives slightly and then springs back into place when released it is almost certainly mounted on anti-vibration mountings. Watch also to see that the fan rocks gently a few times. A fan that falls back into place slowly and sullenly will be overdamped or else be on collapsed springs squashed flat by the weight of the fan.

   d) Listen to the fan.
   A fan that makes the same sound outdoors as indoors is almost certainly producing an airborne noise rather than a structure-borne noise.

2. Complain – but be constructive.
   The landlord or flat-manager is not interested in an isolated opinion. If many people complain he will become concerned about the possible difficulty about re-letting his property. If one person complains he will treat it as insignificant and will attempt to soothe the complainant with sweet words and do nothing about the noise.

8.5   CONTROL OF NOISE IN DOMESTIC CENTRAL HEATING SYSTEMS

There are three principal noise sources associated with such systems: boiler-flues, pumps and thermal expansion. Your own boiler noise is normally acceptable. It has high subjective acceptability in that you know that the noise is caused by a device which is keeping you warm. Boiler noises from other people's central heating systems are another matter.

There are few remedies for boiler-flue noise. If a system is not working correctly a service by a competent service engineer from the boiler manufacturer can often reduce the noise quite noticeably. In rare cases thermo-acoustic resonances can be set up in the chimney or flue which can cause howling and these are a genuine fault. They are caused by organ pipe resonances within the chimney or flue having a positive feedback link to the flame to give a most potent noise source. This can often be cured by bending or breaking up the continuity of the chimney. Often the chimney or flue is a pre-fabricated flexible tube which can be bent but not broken. A remedy here is to place a coil of metal in the form of a spring of at least 25mm cross-section down the chimney to damp the resonances. Trial and error establishes the point at which the resonances can be extinguished.

A pump is often most noisy when it is not an integral part of the boiler but is mounted in a loft space. Experiments have shown that by breaking the pipe on both sides of the pump and by putting in approximately 30 cm of flexible tube connected to the ends of the existing pipe by means of jubilee clips, the noise can be reduced by about 15 dB. Care must be taken in the selection of the flexible pipe especially since a lot of materials are subject to perishing under the effects of hot water. We recommend that this pipe be replaced at least every two years, otherwise a house will be liable to flooding. Also when this is done the pump should be supported by at least 50 mm of fibreglass quilt to stop direct vibration transmission into the building structure.

The noise due to thermal expansion (clicking) usually occurs on a longish run of pipe which is firmly mounted at a number of points along its length. When the system switches on and there is a sudden rush of hot water, the pipe expands and large forces are created between the supports. These forces are released suddenly and explosively in the manner of a minor earthquake causing a very loud click. Often a series of these are heard. The cure here is to release the pipe from most of its supports. Where the supports are necessary for the stability of the pipe they can be replaced by hanging wire mountings or by the existing mountings carefully packed with flexible rubber or fibreglass. The use of polyurethane is not advisable here. By careful work and trial and error, the average 'Do It Yourself' noise-control enthusiast can produce significant improvements and it is not impossible to eliminate this noise altogether.

8.6 COHERENT NOISES – SPECIAL PROBLEMS

Coherent noises usually have a regular waveform and therefore sound tonal in character. It is possible for broad-band or 'white' noises to be coherent but this does not often occur in practice. Two coherent sounds must be combined by adding their pressures, not powers, and the acoustic pressure contains a phase component. If however, the two phases are totally opposed, the pressure fluctuations of the two will sum to zero at all times and silence will be the result. It is not possible to have two coherent noises from different sources cancel each other out at all points in space. The processes of cancelling sound by another sound of opposite phase is called active quietening. But active quietening of a volume of space such as a room is impossible except over volumes whose every linear dimension is much smaller than the wavelength of the sound in question. Since the wavelengths of sound extend from over 3 m at 100 Hz to 3 cm at 10 kHz, we can see that the low frequency "active" quietening will be effective over a larger volume. Attempts to use active quietening have so far been effective in specialised applications – such as the addition of out-of-phase sounds at the ear by means of earphones. A microphone near the ear detects the sound and an electronic device inverts the sound and cancels the pressure fluctuations at the ear. Considerable attenuations have been found possible and the system has the advantage of allowing messages through the earphones to the wearer of the device. This system has applications to flight deck personnel and others who need to communicate in a noisy environment.

Some attempt has been made to prevent walls transmitting sound by coupling vibration-measuring devices and shakers to the wall so as to oppose the movement of the wall and so to stop it reradiating sound falling on it.

The most common example of coherence effects in practice is the occurrence of standing waves in closed spaces; a steady low frequency hum will vary in loudness at different points in a room. At higher frequencies small movements of the head will produce differences in loudness in a steady tone or whistle. It is doubtful if these effects are of any practical use.

8.7 WATER SUPPLY PIPE NOISE

Water pipe noise comes from flushing of WCs, the refilling of WC and other cisterns and the running of water through pipes and water hammer.

The flushing of WCs is a noise that does not affect most people and little can be done about it except by careful layout of the building. More can be done with the noise from filling systems. Recent developments have shown that improvements can be obtained by fitting valves which remain fully open until the cistern is almost full and then close completely over a small residual range of water depth. This reduces the time of filling and the variation in the type of noise heard during filling. An alternative is to place the closing valve float behind a dam so that the cistern fills at full speed until the water overflows the dam, quickly filling the small residual chamber containing the float and stopping the noise abruptly. In some multi-occupancy dwellings, valves which measure out a large flow of water can be provided; these cut out local cisterns completely but require large bore pipes and a sufficient head of water.

When a pipe runs along a wall, turbulence within the pipe itself and at the tap or valve at the end of the pipe produces noise which can be transmitted to the structure of the building and can cause noise at remote points. European standards for this sort of noise are in preparation and in West Germany standards already exist. This sort of noise is best controlled by design of pipe layout but in existing cases it can be reduced by attention to or replacement of valve and taps at the end of pipe runs and by relocating pipes and pipe supports.

Water hammer occurs when a tap or valve is closed suddenly, thus forcing the water down a length of pipe to stop suddenly. The moving mass of water in a pipe carries a considerable inertia and the force needed to stop the water can be high. In these circumstances a shock wave can run up the pipe causing an audible thump which is radiated over a considerable length of the pipe. Slow closing of taps and valves and, in some cases, pressure release traps can cure this problem.

8.8 CONCLUSION

This chapter has attempted to cover some of the problems of domestic noise with the purpose of helping the individual who has a noise problem. Procedures that can be followed to solve domestic noise nuisance and the technical side of the subject are well understood, but it is often found that poor design, installation and maintenance lead to problems. It is to be hoped that the near future will show progress in this field.

The reduction of a neighbour's noise requires co-operation and good sense. Never forget that we ourselves can be our neighbour's noise problem.

9. Noise at a place of employment

9.1 INTRODUCTION

It is known that prolonged exposure to excessive noise can cause deafness, but there are other reasons why it is desirable to be protected from high noise levels. Personnel who are subjected to noise become fatigued and their attention is distracted, and safety records are not so good in a noisy situation. Increasing use of court actions by organised labour to recover damages for loss of hearing due to noise at work, referred to as 'industrial deafness', is occurring, but these cases are difficult to prove against management expertise. Workers in the USA are protected by the Occupational Safety and Health Act (OSHA) (1970). The British government issued 'A Code of Practice to limit the exposure of employed persons to noise' in 1972 and set up a new Health and Safety Commission with powers to approve such codes and give them legal backing in 1974. Another seven of the nine countries of the EEC possess enabling laws on occupational safety and health which could be used to issue regulations on the prevention of hazardous levels of noise at work. However, only France, Germany, Denmark and the Irish Republic have given legislative backing to standards of occupational noise exposure. There is no doubt that these laws give objective standards against which to assess noise climates within factories.

9.2 THE CODE OF PRACTICE FOR REDUCING THE EXPOSURE OF EMPLOYED PERSONS TO NOISE (1972)

This Code lays down that exposure to noise energy equivalent to that due to a steady level of 90 dB(A) for eight hours is to be the limit of exposure allowed (assuming a five-day working week over a working lifetime). The OSHA limit is the same, although the EPA has proposed lowering this limit to an equivalent noise level of 75 dB(A) for eight hours.

The first point to note here is that an L_{EQ} of 90 dB(A) is not the *upper* limit to the instantaneous noise level that employed persons can be exposed to. This is set at 150 dB(A) in the UK and at various values ranging from 135 dB(A) to 115 dB(A) in other countries. However, the time allowed for exposure to levels exceeding 90 dB(A) is limited. In the UK it is limited so that the total noise dose, equal to the product of sound power and time, does not exceed that equivalent to an eight-hour L_{EQ} of 90 dB(A). For example, at a level of 100 dB(A), the rate of receipt of sound energy is ten times as great as at 90 dB(A). So 100 dB(A) can only be tolerated for a maximum of one-tenth of a working day, or 48

Figure 9.1   Nomogram for calculating noise dose increment

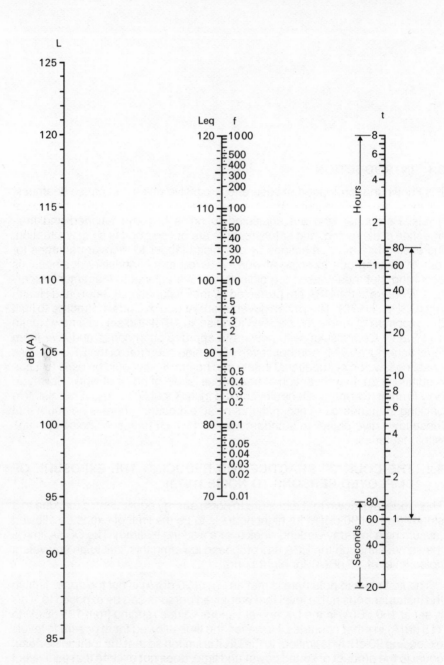

minutes per day. The consequence of this energy rule in the UK is that the allowable exposure period is halved for each 3 dB(A) increase in level. In the USA a 5 dB(A) increase in the level is allowed for each halving of the exposure time. The departure from the energy rule in the USA is to allow for the likely intermittency of exposure during the course of a normal working day.

9.3 ESTIMATION OF NOISE DOSE

There are two ways of estimating the total noise dose of an employed person: the noise survey and work study method, and the noise dose meter.

The first method consists of a survey of the areas of a factory at which noise problems are likely. If it is found that the noise level at no occupied point exceeds 90 dB(A), no immediate further action need be taken. If areas of higher noise levels are found, a complicated set of arithmetic processes must be gone through to determine the total noise dose. In the UK this can be simplified by means of a nomogram published in the Code of Practice and reproduced here (Figure 9.1).

To use the nomogram it is necessary first to prepare a table similar to Table 9.1.

Table 9.1 Calculation of Noise Dose

Noise level dB(A)	Daily exposure time hours/day	Noise dose increment (From nomogram)
	Total dose =	

For each noise level and exposure time, the noise dose increment is read off from line three of the nomogram. All of these exposure increments are added to find the total noise dose. If the total dose exceeds eight hours of 90 dB(A) L_{EQ} the employee is assumed to be exposed to excessive noise. Again, if the total is less than eight hours equivalent no further action need be taken.

It is advisable that factory managers maintain regular records of noise levels, and also check when new equipment is brought into use or when operating schedules are revised. Also, an annual examination of employees' hearing acuity should be carried out to detect unexpected exposures that may have gone unnoticed in a general survey. It is known that between 25 and 30 per cent of the general population are particularly susceptible to hearing damage and would suffer noise-induced hearing loss at lower exposures than the legal limit. These people can be assisted by a screening programme in which an audiologist checks the rate of recovery of hearing from the temporary deafness caused by a short period of exposure to a high noise level.

9.4 THE NOISE DOSE METER

Personnel who must enter factory areas having noise levels in excess of 90 dB(A), for maintenance or periodic testing, can be issued with noise dose meters. Such a meter is analogous to the radiation dose monitor carried by workers in radioactive areas. The dose meter integrates total noise exposure over a period of time and gives a measure of the degree of exposure over that time. This is usually stated in eight-hour L_{EQ} (see Section 2.9). If a figure of 90 dB(A) eight-hour L_{EQ} is exceeded, steps must be taken to reduce the amount of noise to which the employee is exposed.

9.5   CONTROL OF FACTORY NOISE AT SOURCE

1)   *Ordering new equipment*
When new equipment is being ordered and it is expected that a noise hazard may be involved, it should be specified that the noise level at the ears of an operator or minder should not exceed 90 dB(A) as a result of the machine's use. Such action will, if widely used, force plant manufacturers to design and build quieter machinery. The effect will be to increase the capital cost of the plant, but this method of treating the noise problem at source will be the cheapest in the long run.

2)   *Treatment of existing plant*
The most effective means of treatment of noise is at source. However, it is rare to find a case where the immediate source of the noise can be interfered with directly. The most usual sources of high noise levels are pneumatic valves and airjets, boilers, impacting devices such as presses, grinding machines, electric motors, turbines, steam pressure reducing valves and all types of compressors. Very few of these devices can be treated directly once they have been installed and are in use. Indeed any stoppage is likely to be expensive if the machines are part of a production line.

Effective maintenance can count for a great deal. If rotating parts are out of balance, like an unevenly spread load in a spin dryer, and if bearings are excessively worn, the noise level will increase. The sound paths which include transmission through the floor can be broken by mounting the noise source on resilient mountings.

3)   *Machines producing local noise*
If a machine is producing a noise which is locally above the limits but which is not causing the general level to be excessive, the easiest way to deal with this is to see if nearby operators can be moved from the high noise zone. If this is not possible, local noise screening should be tried.

9.6   CONTROL OF FACTORY NOISE ALONG THE PATH

Noise barriers
Noise barriers are a complicated subject, involving, as they do, frequency-dependent diffraction effects which depend on the size and geometry of the barrier. Also it must be remembered that barriers do not eliminate sound energy, but merely reflect and/or diffract it somewhere else. In general a barrier is only reliably effective if its size is equal to or greater than the wavelength of the sound. Thus barriers are more effective at higher frequencies, which have shorter wavelengths. However, a one foot square barrier

placed close to a high frequency noise source, such as an air jet, will reflect substantial amounts of sound energy and protect a nearby operator. Certainly this solution is cheaper than the total enclosing of the source, and does not have some of its disadvantages.

Enclosures

If a machine is producing general high noise levels throughout an area of the factory, it is necessary to consider the possibility of enclosure. Enclosures are frequently impossible for a number of reasons:

1) Operatives must observe and control the process;

2) Materials and products, often large, must arrive and depart from the machine in a regular flow;

3) The thermal insulation of most forms of acoustic barrier will interfere with the process;

4) Ventilation of the process, usually for cooling, may be mandatory;

5) Fire hazards with heat-based processes must be considered.

Some of these difficulties can be overcome in the design of an enclosure and some are intractable.

Where an enclosure is possible it should conform to the following specifications.

The walls should have sufficient mass to provide an adequate transmission loss. Where noise reductions of 20 dB(A) are required thick plywood or blockboard would be adequate. The enclosure must be continuous with no large holes. If it is necessary to have occasional access to a meter or a counter control, a window can be let into the enclosure. Pipes and electricity cables should lead through the enclosure with seals.

It is essential that the inside of the enclosure is lined with sound-absorbing material such as fibreglass. Without sound absorption within the enclosure the sound energy from a machine has nowhere to go so it builds up inside the enclosure until as much sound energy escapes throughout the enclosure as is being produced by the machine. In other words, an enclosure without sound absorption is as bad as no enclosure at all. Of course, there is always some sound absorption present, but this must not be relied upon. The sound absorption prevents the build-up of a reverberant field within the enclosure and as a result the insulating material has only to deal with the direct sound from the machine. Note also that adding a sound absorber to an insulating surface is not necessarily going to increase the sound insulation. Some absorbers, notably polystyrene and cavity (Helmholtz resonator) type absorbers, can reduce the sound insulation of an enclosure.

Where it is necessary to introduce material into the machine or when some ventilation is required, lined ducts should be used. These should be of sufficient length to give insulation equivalent to that of the enclosure. Design charts for the attenuation of ducts are available (see Bibliography).

Moving outward from the source of the noise, we now consider what can be done between the source and the receiver – the nearby employee. None of the above techniques can help an employee close to a noise source since the direct field is dominant and must be reduced before it reaches him.

Noise refuges

The various codes of practice and legal constraints limit the time to be spent at a given level, so that employees do not spend a whole eight-hour period at such levels. Most employees will spend some time during the day at meal breaks, coffee breaks or waiting while production is halted for one reason or another, and if these things happen on a reasonably regular basis the provision of a noise refuge will be useful. This need not be a specially-built structure but merely a nearby room where the factory-produced noise level does not exceed 85 dB(A), below which level exposure time does not count towards the total day's exposure.

Employees should be given instruction to use these refuges for off-time activities, paperwork, etc as much as possible.

Noise screens

In large areas where many processes are occurring it may not be practical to provide refuges, but a form of noise screen may be employed which can limit the spread of noise and provide quiet areas at places of sedentary work. Such screens should be adaptable, as production methods change rapidly in many industries and the provision of fixed noise screens will hamper development. They should not interfere with supervision and control, and they should not get in the way of operators who may be on piece work.

Portable noise screens

A form of portable noise screen that has proved effective in our experience has the following specification.

Panels measuring 2m by up to 1m should be constructed with a wooden frame. The frame should be filled in centrally with a plywood screen at least 6mm in thickness. This should be faced on both sides with at least 25mm of rock wool or glass fibre and the whole covered in hessian or similar open weave fabric. Four of these panels should be hinged at the edges so that the whole screen can be folded flat, or be free standing in a zig-zag manner, or be stretched out straight. These screens are suitable for large areas where the noise level is in the region of 90 – 100 dB(A) and there is no fire risk. A number of these screens should be introduced into the area and instruction given to employees that they are to set them up as they wish in order to reduce noise. Intelligent employees can then distribute them around the area so as to provide noise reduction without impeding work. Of course, some groups may simply proceed to stack them up against a wall. In this case management should point out to a union representative that safety precautions are being neglected by the men and the company might be able to avoid liability for noise deafness. Although under common law in the UK it is essential for the injured employee to show employer's negligence, it would be a reasonable defence to show that all reasonable steps had been taken to reduce the employees' exposure to noise.

9.7 RECEIVER POSITIONS AND NOISE CONTROL

The free field radius

Close to any noise source there is a region where the direct sound from the source dominates over the reverberant field in a room. Very close to the source there is a region where the sound field cannot be directly related to the overall sound power output. This is called the 'near field'. As the distance from the source is increased, we generally enter a region where the sound falls off by 6 dB(A) for each doubling of the distance from the source. This is called the 'free field' and as the level of the free field falls away we reach the point where the free field level is equal to the reverberant field level. The distance of this point, or spheroid in three dimensions, is called the 'free field radius'. For non-directional sources in the same frequency band this radius is controlled by the size and absorption of the room and the level of predominant noise sources in the room. For a dominant noise source the free field radius is independent of the sound energy produced by the dominant source itself.

The effect of adding sound absorption to a room is to lower the reverberant field level and so to increase the free field radius. Any point that is within the free field radius receives almost all its sound energy directly from the source. In small, well-furnished rooms such as small offices and living rooms, since the furnishings themselves produce quite a lot of sound absorption, it is quite possible for the free field radius to exceed the typical dimensions of the room, so that no control over sound levels can be affected by adding more sound-absorbing material to the room.

Measuring free field radius
The free field radius can be measured with reasonable accuracy by placing a sound level meter close to a non-directional small source of white noise* or broad-band noise and then moving away from the source. The measured sound level will be seen at first to fall by up to 6 dB(A) per doubling of distance and eventually reach a steady value. This is the reverberant field level and the distance from the source at which this level is reached is the free field radius.

At a distance from any source there is usually found a uniform background noise caused by all sources filling the room with what is called the reverberant sound field. If the level of this field exceeds 90 dB(A) significantly so that portable screens are not adequate, or if such screens are undesirable for any other reason, the problem is a different one. Normal procedure is to try to identify the dominant noise and to treat this noise at source. When all tractable sources have been dealt with, the reverberant field has to be tackled.

Sound absorption
The use of sound absorbers is effective near high intensity sources because sound absorbers absorb a fixed fraction of the sound energy falling on them. Thus in high intensity situations a lot of energy can be extracted. However in reverberant fields, the halving of the total sound energy in a room means a 3 dB(A) reduction in the sound level. A reduction of 3 dB(A), although it represents a halving of the energy density, is really only just noticeable to the human ear. In fact a reduction of 10 dB(A) is required before the average individual will perceive a halving of subjective loudness and a 10 dB(A) reduction requires that the existing sound absorption be increased by a factor of ten.

The existing sound absorption
As we have remarked, the existing sound absorption in a room can be quite

* White noise contains the same sound level energy at every possible frequency in the audio-frequency range. Any broad-band source contains a wide range of frequencies, eg a steam vent. For the simple determination of the free-field radius any broad-band source would be sufficient.

high. Soft furnishings, carpets, human beings, windows, resonant panels and so on all contribute to the sound absorption and so does the air within the room. Thus the existing area of sound absorption in even an untreated room can be equivalent to enough commercial sound absorbent such as rockwool to cover one wall completely. Thus it can be seen that treatment of the room even on all four walls will produce a maximum of 7 dB(A) of sound reduction. This is hardly cost-effective and of course is based on erroneous ideas, since if too much sound absorption is present the reverberant field itself may cease to exist and the noise level will be controlled by the sound coming directly from the source which cannot be controlled by sound absorption placed on walls. This is controlled by the size of the free field radius.

9.8 REDUCTION OF NOISE AT WORK: ACTIONS TO BE TAKEN BY EMPLOYERS

The responsibility of manufacturers for their employees has risen steadily during the last century and there is no reason to suppose that the trend will not continue. In many countries there are laws controlling noise in the workplace. The present UK code of practice concerning noise in factories will be put into law soon and even today any employer who ignores his responsibility is leaving himself open to possible damage action by his employees individually or collectively. It is essential that any prudent factory manager have a survey of noise levels in his factory carried out at once.

Identification of noise control areas
An initial survey by the employers of the premises will identify all areas which are likely to exceed the 8-hour dose limit of 90 or 85 dB(A). All these areas must be indicated with suitable warning notices and entry should be restricted to authorized personnel.

Ear protection should be provided and further steps taken to compute the total noise dose of persons entering the affected area. In addition to a steady level equivalent to 90 or 85 dB(A), it should be remembered that at no time should an employee receive an impulse exceeding 135 dB (fast response) at his ears or 150 dB (fast response) at any other part of his body. Nor should an impulse greater than 150 dB be received at any time. (Note: these levels are *not* specified in dB(A)).

Training of staff
All staff should be educated in the need to conserve hearing by avoiding noise, using ear defenders correctly, and reporting faulty ear defenders and unusually noisy situations. The union representatives should be consulted on this if any disciplinary action is contemplated. Discipline will be necessary to avoid claims for liability for noise-induced hearing loss from employees.

In medium to large-size works, a senior engineer or safety officer should be given the responsibility of maintaining noise-safety standards within the factory. Such an individual can be sent on a short course concerning noise control techniques and instrumentation so as to avoid the necessity of employing expensive private noise control consultants on simple problems. Many problems, however, are not simple and it is useful to have an engineer on the staff who can suggest when the problem requires the advice of noise consultants, communicate with them and implement their recommendations, and also report on the effectiveness of the consultants themselves.

Ear Defenders
It is important to check that ear defenders are adequate for the purpose. To do this, a statement from the manufacturer of the expected noise reduction in each octave band for the ear defenders should be obtained. This is sometimes difficult in the case of passive amplitude dependent ear defenders, such as ones used by the military to protect against gun noise, and active defenders incorporating electronic systems that pass low-amplitude sounds such as speech and have amplitude-limiting amplifiers that do not pass loud sounds. By referring to the Code of Practice for the reduction of the exposure of employed persons to noise (see Section 9.2) information can be found which can be used to calculate the effective exposure of employees.

In the USA responsibility for certification of earmuffs lies with the Department of Labour who sub-contract the work to private laboratories.

It must not be forgotten that ear defenders are a temporary measure and should not be used in the long term unless it really is impossible to reduce the noise dose of employees below the limit.

Other measures
Planning of new machines and processes should include steps to reduce noise exposure even if this means choosing an entirely different process than the one originally envisaged. It is sometimes possible to schedule the working day so as to limit the noise dose of individuals and to arrange, for example, that short infrequent tasks that involve high noise doses should be performed by staff who usually work in low risk areas (below 85 dB(A)).

Financial aspects
Actions for noise deafness are increasing and out-of-court settlements of such cases can be cited. It is clear that employers who today subject their employees to high noise levels will find shortly either that they will be forced to limit these noises or they will be sued for expensive damages by their employees. Employers in the UK are required by statute to insure against their liability for any personal injury or disease sustained by their employees, arising out of and in the course of their employment. The employer is under a duty to disclose all the material facts facts which he knows to the insurer, such as for example, the number of accidents involving hearing loss. Misrepresentation or failure to disclose any such fact can make the insurance void. The Health and Safety at Work Act enables prosecution of an employer in a criminal court for any breach of his statutory duty to provide a working environment for his employees – that is, so far as is reasonably practical, safe, and without risk to health. A successful claim by an employee for damages for breach of the employer's statutory duty could be met by adequate insurance, but at the risk of an increased premium on renewal. Another liability for an employer arises under section 3 of the Act for injury to persons not in his employment who may be affected by plant on his premises.

An employer may also be liable in common law for injury to his employee's hearing, if the employee can prove that the injury was caused by his employer's negligence. Indeed several successful actions have been fought on this basis.

The Industrial Injuries Act (1965), part of the state social security system in the UK, now provides for compensation of employees in the metal finishing, ship-building and repairing industries for occupational deafness. However the conditions for compensation are extraordinarily stringent.

It is necessary for the claimant

1) to have completed a minimum of twenty years employment in one or more of the appropriate industries which include the use of pneumatic tools,

2) to claim within one year of leaving employment, and

3) to show evidence of substantial hearing loss – an average of 50 dB over the speech frequencies (500 Hz, 1000 Hz and 2000 Hz).

These conditions severely restrict the number of claims that are possible.

The reason advanced by the Industrial Injuries Advisory Council for limiting the range of the scheme and not including other very noisy industries (such as the textile industry, for example) is that existing audiological services are insufficient to deal with the very large number of potential claims.

There is no federal compensation law in the USA. However several states, including Pennsylvania and Georgia, recognise noise-induced hearing loss as industrial accident and thus claims for damages are possible under the accident law. Typically the compensatable hearing loss must equal or exceed 25 dB averaged over the speech frequencies and might be assessed as 1 per cent of the basic salary of the injured employee for each dB hearing loss above 25 dB.

The Canadian Workmen's Compensation System involves collective liability for noise-induced hearing loss on the part of employers who are required to contribute towards insurance in the form of a State Accident Fund. The cost of this is usually passed on to consumers. However, compensation is available to those suffering from at least 25 dB hearing loss over the speech frequencies as a matter of right.

A similar "no-fault" accident insurance scheme was established in New Zealand in 1974, financed by levies on employers and the self-employed. An injured person has an automatic right to receive compensation from the fund if he suffers an industrial disease, regardless of blame.

Of the EEC countries other than the UK, various compensation schemes operate in Germany, France, Belgium and Denmark. In Germany and France the insurance scheme is financed entirely by contributions from employers, but the contribution rates vary according to the accident rates of the particular industry. In Belgium, compensation is paid for occupational deafness by the social security system. In Denmark compensation is paid by the accident insurance company, where the employer must insure the workers against injury. The French compensation threshold is 35 dB average hearing loss over the speech frequencies. In Sweden where noise-induced hearing loss has been recognised as an industrial injury since 1954 the threshold is 35-40 dB.

Records
Some effort should be made to identify those employees likely to be most at

risk. Particularly important from the employer's point of view is the maintenance of records of such employee's hearing acuity when entering employment so that later claims can be checked and cases where the major part of
measured hearing loss was caused by previous employments can be dismis-
sed. Adequate allowance for the normal process of presbycusis (progressive
deafness occurring normally with age) should also be made and the services
of a qualified hearing specialist should be retained.

The ear and hearing loss

The working of the ear was described in Chapter 2. We will now consider the
various ways in which it can fail, revising the basic description as necessary.
Very little can go wrong with the meatus or outer canal which conducts
pressure waves down the eardrum except that it is quite common for low grade
bacteria and flora to accumulate here causing infections, swelling and temporary deafness. Specialist treatment is required to remove these bacteria as
they are extraordinarily persistent. Such infections often occur during the
course of holidays and can cause temporary deafness and tinnitus – that is,
ringing in the ears.

The eardrum, which is not a flat membrane but has the shape of a cone with
an included angle of 120', is connected to a series of bones which are called
ossicles. The purpose of these bones is to transmit the vibrations to the inner
ear. In effect the ossicles act as a transformer matching the low impedance of
the air to the high impedance of the lymphatic fluid within the inner ear. Any
dysfunction of the bones themselves or rupture of the eardrum can cause
instantaneous and total deafness. A type of rheumatism can affect the ossicles
and cause them to seize up. The eardrum can be ruptured by proximity to an
explosion. The resulting deafness is called 'blast deafness' for this reason. In
some cases skilful surgery can repair such damage, restoring a good proportion of the hearing. Once the vibrations have been transmitted across the
ossicles and into the spiral cortex of the inner ear they excite a membrane
which is stretched throughout the spiral region and is called the basilar membrane. This membrane has connected to it a very large number of nerve cells
which are called hair cells. A simplified description is that each of these nerve
cells has a small hair which sticks out into the fluid contained within the cortex
and is excited by the passing sound waves. Each of the hair cells transmits via
its individual nerve and the aural ganglion to the brain. It must not be forgotten
that the brain itself forms an integral part of this hearing mechanism. Thus
damage to the brain can interfere with hearing ability and certain mental
conditions can also cause deafness even though the physical mechanism of
the ear may be working completely normally. Of course hearing is not as
important for man as for some other animals. The brain of a dolphin for
example can be larger than that of the average man. That part of the dolphin's
brain responsible for the auditory processing is particularly large since dol-
phins rely almost entirely on aural signals for communication and navigation.

The most common and insidious form of deafness is the progressive
degeneration of the hair cells in the basilar membrane. Brief exposure to high
intensity noise impairs the function of these hair cells temporarily. Continuous
exposure to high-intensity noises causes physical damage to these hair cells
and they break off and become inoperative. It is an unfortunate fact that nerve
cells throughout the body cannot regenerate themselves, so that once hair

cells are lost they are gone forever. During the course of a lifetime, hearing is impaired by natural wastage of these cells with ageing, a process which is called presbycusis. However there is also damage to hair cells resulting from exposure to excessive noise-induced hearing loss. It is this process of noise-induced hearing loss which most concerns us here since it is avoidable. It has resulted in damages from negligent employers who subject their employees to levels of sound which can cause damage.

Audiometry

The process of measuring hearing sensitivity is called audiometry. This process checks the hearing of each ear of a subject by comparing his threshold of hearing with that of a 'normal' individual. In pure-tone audiometry, monotone sounds of different frequencies, usually 0.5, 1, 2, 3, 4, and 6 kHz, and intensities are presented systematically to the subject, who is asked to pinpoint the intensities at which the sounds are just audible. A *hearing loss* of 30 dB at 2000 Hz for example, would mean that the threshold of the individual under test is 30 dB above a statistical zero that is held to represent 'normal hearing' at the frequency. Figure 9.2 shows the hearing loss of a man aged 71 who had worked at a sawmill giving high noise exposures.

It is possible to test a worker's susceptibility to hearing loss, which varies widely from individual to individual, by subjecting him to noise sufficient to cause temporary hearing loss and monitoring the recovery rate – the slower the recovery rate, the more susceptible the ear.

It is usual to refer to average hearing loss (sometimes called hearing *level)* over the frequencies 500, 1000 and 2000 Hz when assessing the degree of noise-induced hearing loss an individual has suffered. Audiometric testing should be carried out by qualified medical personnel and it is in the employee's interest to see that an audiogram of his hearing acuity is obtained before he embarks on a noisy occupation so that later measurements can be compared to see if damage has occurred.

9.9 NOISE DOSE

The factor which controls the rate of loss of hearing is thought to be the noise dose received by the individual. In most countries this concept has resulted in the rule that a 3 dB(A) increase in noise level results in a doubling of the rate of damage inflicted on ears when the noise is above the damage threshold. This threshold is thought to be about 85 db(A) for 8 hours per day. The 90 dB(A) for 8 hours widely-adopted as a risk limit is in fact high, and incorporates the principle that the employee should accept some risk while the employer must reduce the higher noise levels. Table 3.1 shows a few sound levels for reference purposes.

Belgium, Italy and Canada are among the countries who, like the USA use a 5 dB(A) doubling rule. Denmark, Sweden and Australia are among those who use an equal energy rule like the UK.

9.10 EMPLOYEES' ACTION AGAINST NOISE AT WORK

Employees should ensure through whatever negotiating machinery they have available that the employer has taken the steps outlined above. They should

Figure 9.2 Audiogram of both ears of a man (aged 71) with a history of high noise exposure in a sawmill

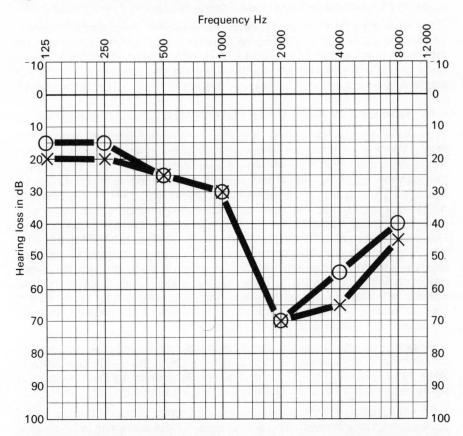

observe rules agreed by management and shop stewards and should always report any changes in noise environment and deterioration of devices or steps taken to ensure that noise is controlled. However, the most important factor here is to realise that ear defenders are important and that they *must* be worn. Physical strength is no guarantee against hearing loss and the penalty for ignoring the use of ear defenders will be increasing deafness, leading sometimes to total loss of hearing in old age. It must be remembered that deafness is a slow process and will not be immediately noticeable. There is no quick loss of hearing, just a slow lowering of the quality of hearing and therefore of the quality of life itself. A (not very reliable) indication of trouble occurs if someone notices a temporary deafness after exposure to noise – the world seems unnaturally quiet. A person experiencing this may well have suffered some small but permanent hearing loss. Tinnitus, ringing or buzzing in the ears, or vertigo after noise exposure, is a definite indication of probable hearing damage and medical advice should be sought.

9.11 NOISE IN THE WORKPLACE AND INSTRUMENTATION

When noise in the workplace seems rather loud to ordinary listening, so that speech is difficult even with raised voices, it is likely that action is necessary according to the Code of Practice (see Section 9.2). The measurement applicable in this circumstance is the Noise Dose (see Chapter 2 and Section 9.3).

For noise control based upon speech interference it is important to be able to analyse the sound into its different frequencies and to determine how much of the energy is in each part of the frequency range. Equipment for this is costly, and if its use is likely to be infrequent it is a good idea to share or borrow it or engage the services of a consultant who has all the equipment at his disposal.

9.12   NOISE AT PLACES OF ENTERTAINMENT

Typically noise levels in places of loud entertainment regularly exceed 100 dB(A). There is a continuing controversy about the effects of such exposure. Some systematic tests have shown a distinct difference in hearing sensitivity between young people (higher or tertiary education students of about 20 years average age) who are regular attenders at places where they are exposed to amplifed popular music and a control group of non-attenders. The same tests have shown a significant increase in hearing loss with rate of attendance. However, the extent of hearing loss, typically 5-10 dB at the speech frequencies, is within the potential measurement error using pure-tone audiometry, and is less than the maximum deviation observed in the tests on the 18-25 age group originally conducted to establish the statistical concept of 'normal' hearing in the UK. Furthermore, the hearing loss revealed by these tests is small compared with the typical industrial noise-induced hearing loss or compared with the typical loss due to gunfire.

Employees in discotheques, and recording engineers are protected against the effects of occupation noise exposure in the UK by the Health and Safety at Work Act (1974) and the *Code of Practice for Reducing the exposure of employed persons to noise* as described earlier in this chapter.

The Association of Professional Recording Studios in the UK has published a paper for its members pointing out the potential risk according to the hearing damage criteria for industrial noise exposure.

Two local authorities in the UK, Leeds and London, have imposed restrictions on the levels of noise in places of loud entertainment: Leeds through its power to lay down conditions on the music licence and London by means of a Code of Practice. Leeds originally laid down a limit on the peak levels, then on the L_{10} noise level and now it is proposed to base a standard on L_{EQ} applicable to all establishments of loud entertainment. The Greater London Council have published a 'Code of Practice for Pop Concerts' covering one-day pop-festivals only. Among the details of stewards and toilets, are proposals for limiting internal noise hazards and external noise nuisance. The proposals regulate the sound levels (L_{EQ}) to 93 dB(A) for an eight-hour period. For shorter periods the level is assessed on an equal energy basis (see Chapter 2). The limit applies to all audience positions at indoor concerts but only at points more than 50 m (164 feet) distant from the loudspeakers at outdoor concerts. Both the Leeds and London authorities are concerned about possible noise nuisance outside the loud music venue. The GLC Code further recommends that during the hours from 7 am to 8 pm the ambient L_{10} level should not be exceeded by more than 5 dB(A) at the facade of the nearest-affected building.

During the night-time hours from 8 pm until 7 am the GLC Code suggests that noise from venues of loud entertainment should not be audible within any potentially affected building.

The Leeds proposals have met with considerable opposition from both regular disco-goers and the music industry. The GLC Code has been found to be difficult to pursue vigorously, both in view of the opposition from the music industry and the considerable expense involved in terms of equipment and manpower to investigate compliance.

There is no scientific consensus that loud entertainment has produced significant hearing loss. However, it seems sensible at least for operators and attenders to take common sense precautions without necessarily reducing the enjoyment obtained. Sound amplification equipment that was designed primarily for outdoor concerts, is now being used at indoor venues, leading to excessive sound output. Industrial experience of such high levels teaches us that there is some risk, particularly to those who are in noisy occupations where they receive the recommended maximum daily dosage before their leisure time.

At outdoor concerts, careful use of audio equipment can reduce the problem of noise outside the event without reducing the sound levels at the concert itself. For example, at a pop festival in Reading, the sound engineer was required to provide a 14 kW public address system without annoying local residents. He achieved this by stacking the loudspeakers high and angling them down into the crowd.

Another possibility for indoor venues is a limitation on the acoustic power output of equipment to be used according to the venue concerned. Generally, the smaller the hall or the closer to residential areas its location, the lower the power output that should be permitted. Noise level switches have been used in the UK, ie devices which cut off the power to the amplifier when a certain pre-set sound level has been exceeded for an agreed length of time. More or less sophisticated warning systems can be built-in. An example is a column of lighted bands which successively light up as the sound level increases. Such devices have operated successfully where the musicians are reasonably sympathetic to the reason behind them, where the pre-set level is reasonable, and where the musicians have learnt to cut down their dynamic range so as to accommodate the noise switch. However, they are not suitable for use wherever there could be the slightest animosity or apprehension about their effects on the part of the musicians or the audience.

In the USA, most communities have not yet done very much to reduce the noise levels in music establishments even though many community leaders continually express the need for a well-organised programme in this sensitive and growing area of concern. Public education through warning signs has been a widely successful method. For example, a typical sign might read:

Warning: Sound Levels Within May Cause Permanent Hearing Impairment

wherever it has been established that loud entertainment levels exceed specific damage risk criteria for the likely exposure period. A community also has the power to limit the sound emitted by the sound system of a place of entertainment through a local noise ordinance.

9.13 CONCLUSION

Despite the existence of relevant legislation or codes of practice there are still many factories and other indoor environments where excessive noise levels are commonplace. Many workers and (more to the point) their unions are not aware of the risk or of the compensation that can be obtained for employers' negligence, wilful or otherwise, in this field. It is essential that anyone who is at all doubtful about the level of noise at his place of work should find out by one means or another how loud the noise is. Simple sound level indicators are quite cheap and if they give readings anywhere near 90 dB(A) serious consideration must be given to noise reduction.

Any neglect of this could lead to permanent hearing damage and expensive damage claims against factory owners.

10.   The economics of noise

10.1   INTRODUCTION

Any noise problem can be solved by current technologies but at a cost. The noise producer and the noise sufferer can benefit from an awareness of the costs of noise control. In this chapter we give some indication of the costs of alleviating noise nuisance and reducing the public health hazards of noise. A different sort of cost is borne by society for every unsolved noise problem and in every situation where noise has some influence on human well-being or behaviour. Many decisions in noise control, and in pollution control generally, relate to the balance between the private or public costs of reducing the level and the social costs of leaving the level of pollution unchanged or the social benefits in monetary terms resulting from the reduced levels. The process of drawing this balance is called cost/benefit analysis (CBA for short).

The important questions are:

1)   How do we quantify the intangible and indirect effects of noise in cash terms?

2)   Do we get a reasonable answer in cash or do we need a new currency?

3)   Is it a good idea to put a monetary value on noise nuisance in the first place?

In the later sections of this chapter we try to answer these questions by referring to examples of cost/benefit analysis with respect to noise.

10.2   COSTS OF REDUCING NOISE INSIDE THE HOME AND THE FACTORY

The increased cost of constructing a sound-insulated building might range from 2 per cent to 10 per cent of the total cost of the building, depending upon geographical area, labour market and other economic factors. With such high cost factors qualified assistance is essential, and probably this is cheapest at the design stage.

Unless quiet is to be a feature of a dwelling and the area in which the dwelling is situated is noisy (near a motorway or dwelling), it is unlikely that money spent on sound insulation will be recouped on sale of the dwelling.

In the case of factories, the noise source is usually internal and can cause cost in two ways, either by subjecting workers to damaging noise or by causing a noise nuisance for nearby residents. The costs of these factors must be weighed against the cost of noise abatement. It is unfortunate that the costs of noise screens, noise-reducing layout partitions, enclosures etc. can be determined easily and are certain to be incurred, whereas the costs of hearing damage actions by employees and from disturbed neighbours are difficult to assess and may not occur.

10.3   COSTS OF REDUCING ROAD TRAFFIC NOISE

At source
It is difficult to cost noise-reduction schemes as data on manufacturing costs is not available outside the vehicle-building industry. However, in considering the maximum possible reduction obtained through engine shielding and fitting special exhausts etc. on a commercial road vehicle, it is possible to arrive at an order of magnitude cost on a weight for weight basis. For example, on a 17.5 tonne gross vehicle weight articulated lorry where the weight of noise treatment totals 40 kg, the noise-reduction measures represent an increase of the order of 0.5 per cent on the running cost per ton-mile of the vehicle.

Road construction
Highway noise can be reduced simply by spending money on the way in which the highway is built. A motorway at grade (that is, at the level of the surrounding land) will create the worst nuisance, followed by elevated, in-cut, cut and cover and tunnel forms of construction in order of decreasing nuisance but in inverse order of cost.

Taking the cost of a motorway at grade with 1.3 m walls running alongside as a guide, and if the resulting value of L_{10} is 5 dB(A) above ambient, a considerable cost is involved to supply 3 m walls and to build the road in a cutting, but such action will only reduce the L_{10} by about 4dB(A). Similarly, earth mounds create an even greater extra cost but again L_{10} is only reduced by about 4 dB(A). The cost of cut and cover is a function of span. However, with suitable flexibility in legislation and the attitude of local authorities, some of the cost can be recouped by making use of the resulting deck for offices and or shops. Cut and cover does considerably reduce the noise from the road.

A particular study of the possible methods of reconstructing the A1 near Hatfield was undertaken to accommodate increased traffic flow without increasing noise nuisance, measured by the number of houses eligible for insulation under the Noise Insulation Regulations 1973. This found that the most effective solution from both the cost and noise points of view was:

1)   to use a cut and cover tunnel,

2)   to displace the motorway to the east of the existing road,

3)   to demolish some flats in the path of the revised route, and

4)   to develop the area on and around the roof of the tunnel with housing units to the value of £2 million (1973 costs)

10.4   COSTS OF REDUCING AIRCRAFT NOISE

An aircraft manufacturer faced with noise certification procedures can choose between a new design of fuselage or wing shape and position so as to shield

112

the ground beneath from engine noise, a new type of 'quiet engine' such as the RB211, or a refit or hush-kitting of an existing aircraft. The costs of new design, of course, are stupendous and even hush-kits can cost up to £4 million per aircraft.

An individual government may elect to reduce noise around its airports by leaning on the airlines to use the quieter types of aircraft or by instituting 'minimum' noise routes and take-off and landing procedures. The latter methods of noise reduction require expensive monitoring and enforcement procedures, and the possibility of increased flight staff payments if safety regulations have to be modified. For the government to cover these costs directly from the public purse or to subsidise airlines indirectly is somewhat inequitable since only a small proportion of the public (those living around the airport) are affected and only a small proportion of the public (the aircraft passengers) are responsible for the nuisance.

The Japanese Ministry of Transport is to surcharge jet aircraft passengers to cover the cost of implementing its noise-abatement programme. The level will be at an average rate of 2 dollars per head but will vary according to the noise level of each type of aircraft using the airports covered by the scheme (including Tokyo, Osaka, Nagoya and Fukuoka). At Frankfurt Airport, in Germany, a different method of covering noise abatement costs is used whereby premiums are awarded to airlines operating with quiet aircraft.

10.5 COST/BENEFIT ANALYSES

Researchers have claimed that factory noise costs the nation one thousand million pounds per annum in lost productivity, illness and inefficiency. However, while we can be fairly confident about the costs of reducing noise levels (the cost of erecting noise barriers, installing double-glazing, fitting hush-kits etc) very little information is available on the cash value of the social or private benefits of reduced levels or about the social or private costs of leaving high noise levels high, so cash estimates are to be viewed with suspicion. Yet if resources are to be saved or correct planning decisions about noise nuisance are to be taken, it is important that we do have some idea of the cost of noise. This is particularly true since

1) standards are often based upon the scientific/political considerations of state intervention, or

2) judgments about permissible levels are left to the vagaries of the judgment of an appointed inspectorate in a planning inquiry, or

3) decisions about the technical or economic feasibility of noise reduction are left to a judge in a law court.

Often the invocation of 'best practicable means' defences operates to protect the noise-maker.

Legislation of some form is necessary since usually so many people are involved that straightforward bargaining is not possible. Taxation or subsidies are traditionally direct ways of achieving a financial balance for a commodity such as peace and quiet. But there is little sign that taxation is to be tried as a means of establishing economically acceptable noise reductions, except in the form of a surcharge on air tickets (Section 10.4). The insulation grant scheme around major airports and insulation according to the Noise Insulation Regulations alongside new or recently modified highways are forms of subsidy. It would be difficult to say, however, that these gave a true indication of the social cost of noise. The maximum amount that people are willing to spend to reduce the noise level in their homes depends very much upon their incomes, while the subsidy is a fixed proportion of the total cost and is not dependent upon income.

Cost/benefit analysis is the most developed method for assessing a technological change or planned development, but it relies upon the identification of *every* cost and *every* benefit, however indirect. Apart from this difficulty, there are problems in finding fair cash relationships: how do we compare cash costs and benefits to a member of the public suffering from noise, which might be of the order of hundreds of pounds, with the costs of noise reduction and benefits of a noisy product or process to the noise producer, which might be of the order of thousands of pounds? Another problem is to decide whether individuals are suffering noise nuisance voluntarily or involuntarily. Finally, there is the all-pervasive problem of placing a cash value upon amenity or freedom from the risk of hearing damage. This last problem is complicated by the fact that not everybody will agree to the same value of amenity or safety even when the value is arrived at by logical steps. Many people feel very strongly about this issue.

The most celebrated and determined attempt in recent times to put a cash value upon noise nuisance was made by the Commission on the Third London Airport, the Roskill Commission. The method of assessing the cost of noise nuisance for each of the proposed airport sites used house prices as an indirect measure. The cost of airport noise was taken to be the sum of:

1)   the fall in freehold property values as a result of noise exposure,

2)   the removal costs and other expenses incurred by those who choose to move to quieter areas as a result of being exposed to aircraft noise, and

3)   the difference between what, in fact, *is* paid for houses in noisy or quiet areas and what *would be* paid by purchasers pushed to the limit.

The result of this calculation for all the proposed sites except one showed that the costs of noise resulting from completion of any of the proposed airports were approximately the same, about one million pounds.

An alternative way of judging the cost of noise from an existing airport is to use an estimate based upon the relative depression of house prices in areas affected by noise greater than 30 NNI around Heathrow and Gatwick. On this basis the devaluation of a £10 000 house is £250 within a 50-60 NNI area, £165 within a 40-50 NNI area and £83 within a 30-40 NNI area. It is simply necessary to add the cost of moving house for each seriously annoyed individual or household.

It has been suggested that if the loss of local amenity expressed in cash terms exceeds one-twentieth of the cash gain, the airport should be shut down. This is because on average one-twentieth of the local population benefit directly from the airport. However, airports are necessary in the absence of an alternative means of really high speed transportation.

10.6 CONCLUSIONS

The choice in planning against noise nuisance from new roads, factories, airports, or neighbours is between a 'prices' approach and an 'authorities' approach. The 'authorities' approach recognises that some people do not have the opportunity or freedom to control the level of noise to which they are exposed. So 'collective', 'social' or 'authoritarian' decisions are laid down about acceptable noise nuisance in the form of 'blanket' standards. Some of the disadvantages of particular 'blanket' standards are spelled out in the next chapter, but the obvious cash disadvantage follows from the fact that a blanket standard might require a noise reduction or a restriction upon planning options which is more than is necessary in view of local circumstances. You can imagine a situation for example where a new road is being built through a community of relatively imperturbable people or where a new estate near an airport is prohibited when it could be filled entirely with airport workers who would not mind the noise of aircraft so much. Where a free market of peace and quiet prevails a 'prices' approach allows more freedom of choice for both the noise-producer and the noise-receiver. But to operate a 'prices' approach at all requires accurate knowledge of the costs of noise reduction and the cost of noise nuisance.

11. How noise legislation is evolved and how standards are set

11.1 INTRODUCTION

It will be mainly through widespread public criticism and awareness of the deficiencies of existing noise legislation that any progress will be made in solving existing noise problems and preventing new ones from arising. In this section we will look first at the existing legislation, the limits that are set and the units upon which they are based. Then we will comment on the general philosphy of legislation against noise control.

11.2 ROAD TRAFFIC NOISE

In the UK the main legislation with regard to control of road traffic noise is the Land Compensation Act, Noise Insulation Regulations (1973 amended 1975). The basis for these is extremely limited. It appears that householders on eleven sites on main roads containing freely-flowing traffic were questioned about how annoyed they were about their environment. All reasons for annoyance, including sleep interference, interruptions to conversations, and vibration, were combined to give an annoyance scale which was compared with the noise level measurements. It was decided that 68 dB(A) was the level at which 95 per cent of the small sample of people scored four out of a possible maximum of seven annoyance points.

As we have already discussed, since a very wide variation in sensitivity to road traffic noise is likely to be found amongst individuals, average annoyance score is not a very satisfactory measure of individual annoyance. It has been suggested that the correlation with individual annoyance can be improved by considering only the most sensitive people; but again, only a relatively small percentage of annoyance can be ascribed to noise level in any case. So the measurement of L_{10} gives only a crude measure of annoyance. Perhaps some improvement can be obtained by separating out daytime and night-time annoyance and correlating these separately with daytime and night-time levels.

The next point is that the use of L_{10}, which is the level that is exceeded for only 10 per cent of the time, might not reflect, or might not take sufficient account of, the disturbance caused by peaks of noise. For example, heavy truck noise will exceed the standard by an indefinite amount. If the average duration of truck noise is 10 seconds, 36 events are possible in an hour, each of which will have some effect on conversation or on concentration or on sleep.

The resulting annoyance may far exceed the mean annoyance upon which the standard is based. Allied with this is the fact that the L_{10} index standard is based upon sites near roads containing freely flowing traffic. There is a need for methods of predicting L_{10} for stop-start traffic, which is more typical on urban roads where noise insulation might be required. Such a prediction method for L_{10} has been derived by a research group at the Building Research Station. They have obtained the relationship:

$$L_{10} = 47.5 + 11.5 \log_{10}(q) + 0.13\,p - 10.3 \log_{10}(D)$$

where

q represents the total traffic flow, p represents the percentage of heavy vehicles and D is the distance from the kerb to the noise-affected facade within a range of 8-23 metres.

This puts more emphasis on the contribution of the percentage of heavy vehicles than the prediction scheme outlined in 'Calculation of Road Traffic Noise'. The formula also indicates that road traffic speed does not contribute to the overall variation in L_{10} for stop-start traffic. A similar conclusion has been reached with the development of a more comprehensive equation by researchers at Imperial College, London, who have found that:

$$L_{10} = 43.51 + 11.23 \log_{10}(q + 8m + 12p) - 0.423c + 4.55 \log_{10} R - 10.21 \log_{10} A$$

where

the extra variables are m the percentage of medium heavy vehicles, c the carriageway width, R is a reflection factor and A is an attenuation factor for nearby surfaces.

The level of 68 dB(A) external to the affected facade is 18 dB(A) higher than the appropriate Wilson Committee night-time recommendation for busy urban areas. The Wilson Committee recommends an internal L_{10} of 35 dB(A) at night and a generous attenuation of 15 dB(A) can be allowed for slightly open windows, giving a total of 50 dB(A) for the external level recommended by the Wilson Committee. Moreover, the same insulation is specified, including double glazing, double doors, mechanical ventilation, sound attenuated ventilation ducts etc., whatever the circumstances. This means that the same insulation has to be applied to a dining room as to a kitchen if they are both on the noise-affected facade and it is clear that this might not be appropriate. It also means that the full attenuation giving up to 35 dB(A), is to be applied however far the noise level is above the 68 dB(A) L_{10} index standard.

Noise insulation has to be awarded only when the level exceeds 68 dB(A) L_{10} index within 15 years of the construction of a new road or modification of an existing one. Yet the main annoyance score on which this standard is based was derived under static circumstances. This means that the 11 sites at which people were questioned were sites where the traffic level was more or less constant for some time. It does not take more than common sense to realise that someone who lives in a rural area or in an area in which the exceeding background level due to traffic is fairly low, will be annoyed when suddenly subjected to a high level of traffic noise, even if the new noise level is less than 68 dB(A) on the L_{10} index. The standard does not really allow for any adjustment for such an abrupt change in circumstances. It is the first year's traffic flow and the first year's noise exposure after the construction or modification that will be the important factor. Perhaps a relative standard specifying the maximum permissible change in noise level would have been more satisfactory.

There are practical difficulties in carrying out the noise monitoring required by the Noise Insulation Regulations. The level to be compared with the 68 dB(A) L_{10} index limit, is the arithmetic mean of eighteen 'hourly' figures (representing the L_{10} within each hour) between 0600 and 2400 hours. There are four methods that can be used to obtain this level:

1) the conventional method with microphone, amplifier, level recorder and statistical distribution analyser, which requires continuous manning with readings taken hourly;

2) a modification of this method in which the equipment is left unattended and the analyser display is photographed at hourly intervals. The lack of manual attention means that untypical noise events near the microphone during the sampled period cannot be allowed for;

3) noise level sampling using a sound level meter and a cassette recorder or directly to an analog-to-digital converter and a data logger, so that the resulting digital data can subsequently be analysed by computer. Most of these alternatives are prone to error and require expensive equipment.

4) the use of modern micro-processor based equipment which can give the 18-hour L_{10} directly but which suffers from the disadvantage of (2) above.

The other legislation concerning road traffic noise is that specified by the Motor Vehicles (Construction and Use) Regulations 1969 amended 1973. Although some successful prosecutions under these Regulations have been claimed in the Huddersfield area (1974), in general the Regulations are difficult to enforce. They require either a special test site or a kerbside check where the road is level and flat, where no adverse climatic or ground cover conditions can be pointed out and where reflection-free measurements can be made at a prescribed distance from the kerb. Such sites are difficult either to construct or to find. The police are not willing to take on the extra load of enforcing these Regulations, probably in view of their commitments and the relatively small penalties incurred by breaking the regulations.

The Regulations themselves are very difficult to read and understand. Different regulations apply to vehicles built before 1st November 1974 to those which apply subsequently. Distinction is made between a kerbside check and a 'production' or 'quality control' check without any meaningful relationship being established between them. For example, the production test requires noise measurements from static vehicles at a specific (high) engine speed. However no reference is made to engine speed in the kerbside test which allows a higher limit even though measurements are made further from the vehicle.

Finally it should be noted that many construction vehicles (particularly

dumper-trucks) that frequently use the highway system and cause considerable noise nuisance are not covered by the Regulations.

11.3   AIRCRAFT NOISE

The Noise and Number Index (NNI) is based upon a survey conducted around
Heathrow Airport and hence upon a certain type of background level, a certain
type of resident and a certain type of aircraft take-off and landing pattern. It is
not surprising that many people consider that the Noise and Number Index
cannot be easily applied to other circumstances, namely to aircraft noise
around Gatwick or Luton or other provincial airports in Great Britain. For
example, it would not always be possible to discount those aircraft which
produce less than 80 PNdB peak level at the point considered. It is also
possible that the relationship between a number of aircraft and their average
peak level and noise annoyance is not correct. Some Swedish work has
shown that within an exposure category which is determined by the frequency
of exposure, the noisiest aircraft determine annoyance. Indeed this is in line
with the type of consideration suggested by the Traffic Noise Index. So the
critical contour, according to the Swedish work, is that of the *noisest regular
aircraft* and is determined by the frequency of that aircraft. Noise from less
noisy aircraft is immaterial. Even if the Noise and Number Index is accepted as
a measure of aircraft noise annoyance around Heathrow, there are problems
in implementing controls and standards based upon it. There are uncertainties
in pilot behaviour and in climatic conditions. A temperature inversion such as
that which occurs on a foggy day or almost every night, causes aircraft noise to
be heard over a much larger area than during days when there is no temperature inversion.

Another criticism that can be levelled at NNI is the lack of explicit account of
the difference between night-time and daytime annoyance. Disturbance of
sleep is perhaps the most annoying effect of noise. The Noise Exposure
Forecast does allow for the difference between night-time and daytime
annoyance. In fact the difference in the constants K (see Chapter 4) in the
expression for NEF suggests that a single night-time flight contributes as
much to NEF as 17 daytime flights. The value of daytime K is chosen arbitrarily
(as is the −80 in the expression for NNI) so that the resulting numbers lie
typically in a range where they are not likely to be confused with other
composite noise ratings.

An overriding objection is the same as that made against traffic noise and
results from the wide variation in individual response. The Noise and Number
Index correlates well only with average annoyance of individuals and not with
the annoyance of each individual.

Essentially, the status of any of the airport noise forecasting techniques is
that of a planning tool. With any such technique a point of noise saturation can
be defined and lower grades of noise contamination may be estimated. The
implementation of any of the techniques for this purpose requires the aid of a
large high-speed digital computer and a very large amount of data describing
the noise characteristics of each type of aircraft. However, a recent study has
suggested a more straightforward relationship between the PNdB limits of
each individual noise event and the number of such noise events per day. For

example, as many as 20 events per day of 90 PNdB can occur without serious problems arising. If, however, the number of noise events exceeds 1000 per day (as it would near a busy airport) then the maximum PNdB of each event that would not create serious problems is reduced to 70 PNdB.

11.4 INDUSTRIAL NOISE (INTERNAL)

The extent to which workers in noisy environments can claim compensation under the Industrial Injuries Act (1965) in the UK leaves much to be desired. In the first place, the range of industries covered by the compensation scheme as far as noise is concerned is extremely small compared with the range for other industrial injuries and the range for noise-induced deafness which is normal in both the United States and Europe. The main argument advanced against extending the scheme is the paucity of existing audiological services. Secondly, this scheme where it does apply is extremely restrictive. The number of workers that can apply for compensation is severely limited by the conditions that are laid down by the scheme. An average hearing level [of 50 dB] across the speech frequencies has to be demonstrated. The worker has to have been in the noisy industry for at least 20 years. The noise-induced deafness has to have been shown to have been incurred during the course of the industrial employment. The main reason why such a severe average hearing level is required is that the criterion of deafness being used is 'difficulty in interpreting everyday speech'. However this does not take account of the host of other disabilities which are consequent upon relatively mild deafness with regard to the hearing of everyday speech. Difficulties in hearing the unfamiliar word or the unexpected warning, or in hearing a new telephone number or message of unfamiliar content, are all criteria that can be used in assessing the degree of handicap due to noise-induced deafness.

Finally, all assessments of noise-induced hearing loss are based upon the British Standard concept of normal hearing. This in turn is based upon measurements carried out in 1967 on approximately 1800 individuals between ages of 18 and 25 who did not have any known history of serious exposure to noise. What was normal hearing in 1967 in this group of individuals could well be different now in the early 1980's among the equivalent groups of individuals. Reasons for this are the increasing exposure to noise in leisure time – for example discotheques, power boats etc, which can lead to sociocusis and an increased risk of damage to hearing outside of one's normal employment. Any scheme to assess noise-induced hearing loss should take this into account.

11.5 INDUSTRIAL NOISE (EXTERNAL)

A major problem with BS 4142 method, of rating the likelihood of annoyance from an industrial source in a mixed residential area, arises when the prevailing background level cannot be measured. In this case (see Chapter 7), it is necessary to estimate the background level from knowledge of the type of area and time of day including corrections for assessing whether the industry is new, and so on. There is evidence that the method of prediction given in this British Standard is in error. It severely underestimates the prevailing background level in areas which already contain industry. As a result, annoyance from new industry in these areas would be severely underestimated. In addition, the standard has problems in dealing with a multiplicity of sources where it is important to identify the particular source causing the annoyance. Further-more the assertion, based upon the Standard, that there would be no annoyance registered in homes where the external background level was as high as 80 dB(A) and where the noise due to industry peaked at 85 dB(A), including all corrections, is clearly absurd.

With the noise-abatement zones specified under the Control of Pollution Act (1974), the main problem for a local authority will be to decide upon acceptable levels to be laid down as registered levels. It might be that central government guidance can be given according to type of industry classification. However, it should be borne in mind that the general load upon local authorities of the type of legislation we have described so far is enormous. They have to monitor many cases where the Land Compensation Act, Noise Insulation Regulations (1973 amended 1975) can be invoked, in some areas they have to consider the problem of entertainment noise and investigate accordingly and they also have the problem of designating noise-abatement zones and enforcing their operation.

11.6 PHILOSOPHIES OF NOISE LEGISLATION AND STANDARDS

None of the multiplicity of noise units that have been derived for different individual sources of unsteady sound has been used to set legally enforceable standards relating to the real-life mixture of sounds. The fact that there are so many noise units results from the difficulties of measuring unsteady noise exposure in terms that can be related to human response. This is really no fault of acousticians, but in making diverse attempts to overcome the difficulties, acousticians have sometimes succeeded in magnifying the problems of environmental control and adding to the confusion of those who wish to solve these problems.

The advantages of the more complicated noise units over the others are not at all clear. Although increasing sophistication in physical measurements can lead to better correlation with average response to noise, the correlation with individual response remains poor. This is not surprising since, as we have explained before, a large part of an individual's response to a noise depends upon 'non-acoustic' factors.

So there seems to be a case for a single noise unit which is easy to understand, to measure and to predict and which can be used to lay down standards relevant to any noise problem. This would be similar, for example, to the Beaufort scale of wind strength or the Richter scale for describing the magnitude of an earthquake. The Noise Advisory Council has recommended the use of L_{EQ} and the Environmental Protection Agency in the USA has advocated the use of L_{EQ} (24-hour) for this purpose. This simple concept of equivalent sound level is somewhat refined to take account of the greater disturbance that is experienced from a certain noise at night than from the same noise by day. Basically the night-time noises are treated as if they were 10 dB noisier than they actually are. The resulting L_{EQ} (24 hours) is known as the day-night sound level or L_{DN}. This unit is used by EPA to set environmental goals (rather than legally-enforceable standards). The implication of setting an L_{DN} of 60 dB(A) as an environmental goal is that no one should be subjected to a sound of 100 dB(A) or above for more than one second in any 24 hours.

Of course the same problem arises with an environmental goal in this form as with any blanket standard – there is always a local circumstance in which a lower or higher limit would be more appropriate. It would seem to be better to have locally-based environmental goals to match local circumstances and, if possible, covering all forms of pollution. Indeed we have examples of both centrally and locally-defined environmental standards in British legislation against noise; the 68 dB(A) L_{10} index of the Land Compensation Act, Noise Insulation Regulations (1973 amended 1975) and the Registered Noise Levels of the Noise-Abatement Zones designated according to the Control of Pollution Act (1974) can both be criticised on different grounds – the former because of its inflexibility and the latter because of the technical and administrative load that it places upon local authorities: how to decide upon limits and how to enforce them. This is probably why the EPA is not suggesting that the goals should be legally enforceable. So finding a unified noise unit alone will not solve noise problems or produce effective noise legislation. The real answer lies in laying down and enforcing emission standards on individual sources. These must be justified in terms of their effectiveness in reducing the total real-life noise exposure from 'real-life' mixtures of sources. This means that a unified noise unit is necessary and important for setting goals from which the individual source restrictions can be derived.

Appendix A

CALCULATION OF L_{EQ}, L_{10} AND OTHER PERCENTILES

L_{EQ} and other statistical units such as L_{10} and L_{90} are not directly and accurately measurable using a simple A-weighted sound level meter. However it is possible to calculate good approximations to these units using such simple measuring equipment. The following is a guide to how this can be done.

Measuring procedure
When measuring traffic noise in the UK it is essential to follow the instructions in the *Motor Vehicle (Construction and Use) Regulations (1969)* or in *Calculation of Road Traffic Noise* (see Chapter 3 and bibliography for more complete reference) as to the setting up and placing of sound level meters. For factories and construction sites the correct procedure is described in DOE Circular 2/76 (loc cit Chapter 6) and for noisy machines the procedure is given in *The Code of Practice for the protection of Employed Persons from Noise* (see Chapter 8). The meter should normally be set to 'fast' for traffic noise and 'slow' for other types of noise including background noise, even if caused by traffic (this will occur when measuring background noise level (L_{90}) for the purposes of assessment using BS 4142 (see Chapter 8)).

Taking readings
To determine L_{EQ} etc it is necessary to accumulate a set of instantaneous measurements at regular time intervals (for example, 100 readings taken at 10s intervals). It is important to ensure that the readings are taken at the precise moment when they are due and that there is no holding back in order to obtain, for example, a higher reading as a lorry passes by. It is permissible, however, to miss a few readings (making up the total at the end) for purely random reasons and it is, of course, essential to miss readings if a louder noise, such as a passing train or aircraft drowns out the object of your measurements. Ideally you should persuade a partner who has an appropriate chronometer to tap you on the shoulder at the correct times and you then write down the reading at once or tick it off on a prepared list. An example of such a list is shown in Table A.1.

L_{EQ} is then calculated from the formula:

$$L_{EQ} = 10 \log \left[\frac{\sum n_i x_i + 60}{\sum n_i} \right]$$

Calculation of L_{EQ}

Readings are taken at equal time intervals and ticked off on Table A.1. This has been chosen arbitrarily to allow a dynamic range of 40 dB from 60 – 100 dB. After the measuring session the total number of ticks at each level (n_i) is found and entered in column 3. The total of all the n_i's is found and entered at the bottom of column 3. This total should not be less than 100. The individual n_i's are then each multiplied by the appropriate correction factor (x_i) which has been calculated from the definition of L_{EQ} and is shown in column 4 to give a set of products $(n_i x_i)$ in column 5. These are totalled at the bottom of column 5.

As long as the dynamic range does not exceed 40 dB(A) then Table A.1. is appropriate. For example, if the measured noise levels vary between 50 and 90 dB(A), simply add 10 dB(A) to all levels. Use Table A.1. as described above and finally subtract 10 dB(A) from the calculated L_{EQ} value. If the lowest measured noise level is 70 dB(A) use Table A.1., and finally subtract 10 dB(A) to the resulting L_{EQ} value, etc. L_{EQ} should be specified as determined over a stated period of time, particularly if the period covers a discrete sound event. For example, if 100 readings are taken at 10 second intervals the calculation would yield an L_{EQ} over 15.67 min. This enables a determination of the total noise dose to be made if necessary.

Calculation of L_{10} and other percentiles

L_{10} is the level exceeded for 10 percent of the time. A good approximation can be obtained by deleting 1/10 of the readings, starting with the largest reading and working down in noise level. If a total of 100 readings has been obtained this means deleting the highest 10 readings. When this has been done the highest remaining reading will represent the L_{10}. L_{90}, as background noise level (for BS 4142 for example) can be similarly obtained by deleting one tenth of the total number of readings starting with the smallest reading.

Table A.1.

Sound level, i (dB(A))	Ticks	No. of Readings, n_i	Correction factor x_i	$x_i n_i$
60			1.0	
61			1.259	
62			1.585	
63			1.995	
64			2.512	
65			3.16	
66			3.98	
67			5.01	
68			6.31	
69			7.94	
70			10	
71			12.59	
72			15.85	

Table A.1.

Sound level, i (dB(A))	Ticks	No. of Readings, n_i	Correction factor x_i	$x_i n_i$
73			19.95	
74			25.12	
75			31.60	
76			39.80	
77			50.10	
78			63.1	
79			79.4	
80			100	
81			125.9	
82			158.5	
83			199.5	
84			251.2	
85			316.0	
86			398.0	
87			501.0	
88			631.0	
89			794.0	
90			1000	
91			1259	
92			1585	
93			1995	
94			2512	
95			3160	
96			3980	
97			5010	
98			6310	
99			7940	
100			10000	
Totals	$\sum n_i$			$\sum x_i n_i$

Appendix B

HOW TO OBTAIN HELP WITH NOISE PROBLEMS

General
In the UK advice and assistance on matters of noise control may be obtained from a variety of organisations including amenity societies, local authorities and government agencies. A similar situation obtains in many other countries. In the following we summarise the system in the UK in the expectation that the equivalent authority or organisation in many other countries can be readily identified. Further details may be found within the main body of the text.

Neighbourhood Noise Nuisance
In the United Kingdom the first line of approach after informal direct contact with the person making the noise is to your Local Environmental Health Department. Such departments exist in each of the local government offices throughout the UK. Often there is more than one individual who can deal with noise and some of the larger metropolitan areas have whole divisions devoted to noise.

The Local Environmental Health Officer (LEHO) is not, however, obliged to help. He will decide on the basis of his experience, a first-hand appraisal, and in the light of other calls upon his time whether or not to pursue any particular noise problem.

In the event that the LEHO decides that a noise is a nuisance, then this noise becomes a statutory nuisance and legal machinery should be invoked to abate the nuisance. This will not necessarily happen, however, since the defence of the best practicable means is available to noise producers who are a business or a person carrying on a business.

In the event that the LEHO decides that a noise is not a nuisance he will take no further action and the next step is to try to use *Section 59 of The Control of Pollution Act (1974)* to abate the nuisance. To do this you have to satisfy a magistrate that the noise is a nuisance. This is not easy since the noise-producers will be acting against you and they will probably have more resources to call on than yourself. You should employ a qualified noise consultant to advise you before proceeding. You should write to The Association of Noise Consultants, 6 Long Lane, London EC1A 9DP to obtain details of consultants available in your area. Your Local Environmental Health Officer or factory inspectorate will also usually know of suitable consultants in your area. Your case will be improved if you can get your neighbours to join in and local action groups responsibly organised will have even better credibility in court. Proceedings under Section 59 will still fail if the defence of the best practicable means is successfully used.

A civil action for nuisance can be taken by property owners and can be pressed without fear of the best practicable means defence since this is not available against such action. Injunctions against the noise and expenses or damages can be obtained this way. However, such actions are expensive and legal advice must be obtained from a solicitor before proceeding.

Many small noise nuisances come under the control of the local authorities and you should ascertain whether a minor nuisance is the subject of a local bye-law and complain to the police before taking other action.

Road traffic noise
Road traffic noise is controlled by the Motor Vehicle (Construction and Use) Regulations 1973, which lay down limits for individual vehicles, and noisy vehicles should be reported to the police. Noisy roads, however, should be brought to the attention of the local highway authority. Noise from new roads or recently modified roads may be considered under the terms of The Land Compensation Act, Noise Insulation Regulations (1975). Local authorities are obliged to administer compensation or insulation schemes in these cases and early application to them is essential.

Aircraft and airport noise
Aircraft and airport noise is the responsibility of the Department of Trade and the Civil Aviation Authority. Complaints should be made to the operator of your local airport. Most aircraft noise, however, is excluded by statute from the body of actionable noise nuisance. In the case of military aircraft, complaints should be made to the station commanding officer and then to the Ministry of Defence, Provost and Security Services UK, Government Building, Bromyard Avenue, Acton, London W3.

Construction and demolition site noise
According to Sections 60 and 61 of The Control of Pollution Act (1974), local authorities, again through the Environmental Health Department, can control this sort of noise. Sometimes, however, the Local Authority will give construction companies permission to create a certain amount of noise at prescribed times. This does not preclude Section 59 action by individuals but will make success most unlikely as the magistrate will be aware of the existence of such permission. In fact the defence should make the magistrate so aware.

Noise in the workplace
Matters relating to the risk of hearing damage are the proper responsibility of a Safety Officer in the employ of any sizeable industry. He or she will be aware of relevant legislation in The Health and Safety at Work Act (1974) administered through the Health and Safety Executive. A local Medical Officer will also be a source of advice.

Other sources of help
The Noise Abatement Society, 6 Old Bond Street, London W1X 3TA exists to co-ordinate action and advice on noise problems but in practice serves mainly to collect and distribute press reports on noise matters.

The Noise Advisory Council is a government body which has produced many useful booklets on relevant terminology and legislation. These are available from HMSO and provide authoritative information on noise problems.

The Building Research Establishment, 1, Bucknells Lane, Garston, Watford WD2 75R, Herts, offers an advisory service to local authorities, contractors and industry on matters of sound insulation and noise nuisance. A calibration facility for measuring instrumentation and related technical advice is available at The National Physical Laboratory, Teddington, Middx TW11 0LW.

The Institute of Acoustics, 25 Chambers Street, Edinburgh EH1 1HU is a source of information on careers and education relating to noise control.

References and bibliography

CHAPTER 1

1.1 Committee on the Problem of Noise (1963) Noise, Final Report, HMSO, London.

1.2 Ministry of Transport Steering Group and Working Group; On the Study of Long Term Problems of Traffic in Towns. (1963): a study of the long term problems of traffic in urban areas; Reports of the steering group and working group appointed by the Minister of Transport (Chairman: Colin Buchanan), HMSO, London.

1.3 Greater London Council (1970) Urban Design Bulletin: Traffic Noise, GLC, London.

CHAPTER 2

2.1 Burns, W. (1972) Noise and Man, John Murray, London.

2.2 Kryter, K. D. (1970) The effects of Noise on Man, Academic Press, New York.

2.3 Chedd, G. (1971) Sound, Aldus Books, London.

2.4 Stevens, S. S. (1966) Sound and Hearing, Time Life International, New York.

2.5 BS 3383 (1961) Normal Equal Loudness Contours for Pure Tones and Normal Threshold of Hearing under Free-field Listening Conditions, British Standards Institution, London.

2.6 BS 661 (1969) Glossary of Acoustical Terms, British Standards Institution, London.

2.7 Noise Advisory Council (1975) A Guide to Noise Units, HMSO, London.

CHAPTER 3

3.1 Doelle, L. L. (1972) Environmental Acoustics, McGraw-Hill, New York.

3.2 Yerges, L. F. (1971) Sound, Noise and Vibration Control, McGraw-Hill, New York.

3.3 Kinsler, L. E and Frey, A. R. (1962) Fundamentals of Acoustics, second edition, John Wiley, New York.

CHAPTER 4

4.1 Chalupnik, J. (Ed) (1970) Proceedings of the Washington Symposium on Transportation Noise, Washington University Press, Washington D.C.

4.2 Noise Advisory Council (1972) Traffic Noise: the vehicle regulations and their enforcement, HMSO, London.

4.3 BS 3425 (1966) amended 1973, Measurement of Noise Emitted by Motor Vehicles, British Standards Institution, London.

4.4 Organisation for Economic Co-operation and Development (1971) Urban Traffic Noise, OECD, Brussels.

4.5 Department of the Environment and the Welsh Office (1975) Calculation of Road Traffic Noise, HMSO, London.

4.6 Road Research Laboratory (1970) A review of road traffic noise, RRL Report LR 257.

4.7 Department of the Environment and Control Office of Information (1973) Land Compensation – your rights explained, Booklets 2 and 5, HMSO, London.

4.8 OECD (1970) Urban Traffic Noise, OECD, Paris.

4.9 Legg, J. D. (1980) Urban Railway Noise and New Housing, OECD, Proc. Institute of Acoustics Spring Meeting pp. 81-84.

4.10 Hemsworth, B. (1980) Prediction of Noise from Trains, Proc. Institute of Acoustics Spring Meeting pp. 89-92.

CHAPTER 5

5.1 McKennel, A. et al (1971) Second Survey of Aircraft Noise Annoyance around London (Heathrow) Airport, HMSO, London

5.2 Noise Advisory Council (1972) Aircraft Noise: Selection of Runway Site for Maplin, HMSO, London.

5.3 Department of Trade and Industry (1972) Action against aircraft noise DTI, London.

5.4 Department of the Environment (1973) Planning and Noise Circular 2/73, DOE, London.

5.5 Noise Advisory Council (1974) Review of Aircraft Routing Policy, HMSO, London.

CHAPTER 6

6.1 BS 5228 (1975) Code of practice for Noise Control on Construction and Demolition Sites, British Standards Institution, London.

6.2 U.S. Department of Housing and Urban Development (1971) Environmental Noise Standards for Construction Sites, HUD circular 1390-2, Washington D.C.

6.3   Department of the Environment (1971) Planning and Noise Circular 10/73 (16/73 Welsh Office), HMSO, London.

6.4   Department of the Environment (1972) Noise Control on Construction Sites Advisory Leaflet, HMSO, London.

CHAPTER 7

7.1   BS 4142 (1967 amended 1975) Method of Rating Industrial Noise Affecting Mixed Residential and Industrial Areas, British Standards Institution, London.

7.2   Noise Advisory Council (1972) Neighbourhood Noise, HMSO, London.

CHAPTER 8

8.1   Noise Abatement Society (1970) The Law on Noise, Noise Abatement Society, London.

CHAPTER 9

9.1   Department of Employment (1971) Noise and the Worker, HMSO, London.

9.2   Department of Employment (1972) Code of Practice for Reducing the Exposure of Employed Persons to Noise, HMSO, London.

9.3   Beranek, L. L. (1972) Noise and Vibration Control, McGraw-Hill, Englewood Cliffs, N.J.

9.4   Magrab, E (1978) Environmental Noise Control, Edward Arnold, London.

9.5   Hay, B (1975) Occupational Noise Exposure – The Laws in E.E.C. Sweden, Norway, Australia, Canada and the U.S.A., Applied Acoustics pp. 299 – 314.

CHAPTER 10

10.1   Waller, R. A. (1967) Environmental Quality, its measurement and control. International Seminar on Urban Renewal, Brussels.

10.2   Roskill Commission (1970) Commission on the Third London Airport: Papers and Proceedings Volume VII – parts I and II, HMSO, London.

10.3   Attenborough K., Pollitt, C. and Porteous A. (1977) Pollution: The Professionals and the Public, Open University Press, Stony Stratford, UK.

CHAPTER 11

11.1   Bryan, M. (1973) Noise Laws: Don't Protect the Sensitive, New Scientist, 27 September.

11.2   Attenborough K. (1976) Why our noise laws are inadequate. New Scientist and Science Journal, 6 May.

11.3   Mulholland, K. A. and Attenborough, K (1971) Predicting the noise of Airports, New Scientist and Science Journal, 18 March.

11.4 Osborn, W. C. (1973) The Law Relating to the Regulation of Noise Chapter 11 of P. McKnight, P. Marstrand and C. Sinclair (eds.) Environmental Pollution Control, Allen and Unwin, London.

11.5 Gilbert, D. Moore, L and Simpson, S (1980) Noise and Urban Traffic (Interrupted Flow), Proc. Institute of Acoustics Spring Meeting pp. 113 – 114b.

11.6 Noise Advisory Council (1978) A guide to the measurement and prediction of the equivalent continuous sound level L_{EQ}, HMSO, London.

A. BIBLIOGRAPHY – GENERAL

A.1 BBC (1971) Quieter Living, BBC Publications, London.

A.2 Taylor, R. M. (1975) Noise, Penguin Books (Pelican), Harmondsworth.

A.3 Penn, C. (1979) Noise Control, Shaw and Sons Ltd., London.

A.4 Milne, A. (1979) Noise Pollution: Impact and Counter Measures, David and Charles, London.

A.5 Barden, R. G. (1976) Sound Pollution, University of Queensland Press, Auckland.

A.6 Rodda, M. (1967) Noise and Society, Oliver and Boyd, Edinburgh.

A.7 Cunniff, P. F. (1967) Environmental Noise Pollution, John Wiley and Sons, New York.

A.8 Kerse, L. A. (1974) Noise and the Law, Oyez Publications, London.

B. BIBLIOGRAPHY – TECHNICAL

B.1 Handbook of Noise and Vibration Control (1974) edited by R. A. Warring, Trade and Technical Press.

B.2 Noise Reduction (1972) Leo. L. Beranek, McGraw-Hill, Englewood Cliffs, N.J.

B.3 Handbook of Noise Control (1957) edited by C. M. Harris, McGraw-Hill, Englewood Cliffs, N.J.

B.4 Acoustics and Vibration Physics (1966) R. W. B. Stephens and A. E. Bate, Edward Arnold, London.

B.5 Teach Yourself Acoustics (1967) Jones, G. R. Hempstock, T. I. Mulholland, K. A. and Stott, M. A., English Universities Press, London.

B.6 Woods Handbook of Noise Control (1972) Sharland, I., Woods of Colchester Ltd., Colchester.

B.7 Environmental Noise Control (1975) Magrab, E. B., John Wiley & Sons, New York.

B.8 Economic Analysis of Transportation Noise Abatement (1978) Nelson, J. P. Ballinger Publishing Company, Cambridge, Massachusetts.

Index